Lecture Notes in Economics and Mathematical Systems

Managing Editors: M. Beckmann and W. Krelle

358

Paul Knottnerus

Linear Models with Correlated Disturbances

Springer-Verlag
Berlin Heidelberg New York London Paris
Tokyo Hong Kong Barcelona Budapest

Author
Dr. Paul Knottnerus
PTT Telecom, I & AT
P.O. Box 423
NL-2260 AK Leidschendam
The Netherlands

ISBN 3-540-53901-8 Springer-Verlag Berlin Heidelberg New York
ISBN 0-387-53901-8 Springer-Verlag New York Berlin Heidelberg

Printed in Germany

Printing and binding: Druckhaus Beltz, Hemsbach/Bergstr.
2142/3140-543210 – Printed on acid-free paper

ACKNOWLEDGEMENTS

Herewith I would like to express my gratitude to my supervisor Professor H. Neudecker for his guidance, encouragement, humour and serious enthusiasm which have led to this volume. Apart from some minor revisions the text is identical with the author's Ph.D. thesis, "Linear Models with Correlated Disturbances," University of Amsterdam.

I am also deeply indebted to Professors J.G. de Gooijer, F.C. Palm, G.D.A. Phillips, T.J. Wansbeek and to Drs. I.J. Steyn for their helpful comments on an earlier draft of various chapters; they often drew my attention to specific literature, I was unaware of at the time.

Special thanks must go to Professor J.F. Kiviet for his insistence and encouragement. Moreover, I gratefully acknowledge his many helpful comments after a thorough reading of a first draft of the manuscript.

I am also grateful to the many authors whose books helped me understand Econometrics, especially Professor A.C. Harvey as author of "The Econometric Analysis of Time Series", and I ask the forgiveness of those authors whose name I have not always mentioned in the text in spite of their influence.

Furthermore, I wish to thank the many friends and colleagues for fruitful discussions, invaluable suggestions and encouragement during the last couple of years. In particular, I am indebted to Jan Weide for detecting a number of inaccuracies in the final draft. Remaining inaccuracies and errors are, of course, my sole responsibility.

Finally, I would like to thank PTT Telecom for providing me with the office facilities and last, but by no means least, I am very much obliged to Yvonne van Elswijk-Bremer for typing such a difficult manuscript.

CONTENTS

CHAPTER I

INTRODUCTION

In each chapter of this volume some specific topics in the econometric analysis of time series data are studied. All topics have in common the statistical inference in linear models with correlated disturbances. The main aim of the study is to give a survey of new and old estimation techniques for regression models with disturbances that follow an autoregressive-moving average process. In the final chapter also several test strategies for discriminating between various types of autocorrelation are discussed. In nearly all chapters it is demonstrated how useful the simple geometric interpretation of the well-known ordinary least squares (OLS) method is. By applying these geometric concepts to linear spaces spanned by scalar stochastic variables, it emerges that well-known as well as new results can be derived in a simple geometric manner, sometimes without the limiting restrictions of the usual derivations, e.g., the conditional normal distribution, the Kalman filter equations and the Cramér-Rao inequality.

The outline of the book is as follows. In Chapter 2 attention is paid to a generalization of the well-known first order autocorrelation transformation of a linear regression model with disturbances that follow a first order Markov scheme. Firstly, the appropriate lower triangular transformation matrix is derived for the case that the disturbances follow a moving average process of order q (MA(q)). It turns out that the calculations can be carried out either analytically or in a recursive manner. The derivation is based on a specific Euclidean linear vector space. Besides results from the theory of difference equations are used. Using the same approach it also appears to be possible to derive an exact expression for the inverse of the autocovariance matrix of an MA(1) process, which was first found by Arato (1961). Secondly, the appropriate transformations are derived for the case that the disturbances follow an arbitrary mixed autoregressive-moving average process of orders p and q, respectively, denoted by ARMA(p,q). Also a formula is derived for the autocovariance matrix of a general ARMA(p,q) process. Although several algorithms have been published for the calculation of the autocovariance matrix, the derivation given here, has to my best knowledge not been published elsewhere; see McLeod (1975), Ansley (1980) and Kohn and Ansley (1982). Zinde-Walsh (1988) gives a somewhat different expression for the autocovariance matrix of an ARMA(p,q) process. Since maximum likelihood estimation requires the Jacobian of the transformation, expressions for the Jacobian are derived as well.

In Chapter 3 it is shown how the calculations of the appropriate transformations described in Chapter 2, can be carried out in such a way that the number of multiplications is of order n, n being the number of observations. Also it is shown that the approach of Ansley (1979) for calculating the likelihood function of a general ARMA(p,q) process can be adopted within the theoretical framework of Chapter 2. This gives some more insight in Ansley's article which is not totally self-contained and so it refers to several other authors for efficient algorithms. In our view this is somewhat unsatisfactory. Chapter 3 is also intended to shed some light on Ansley's approach from a more geometric viewpoint.

In Chapter 4 the estimation method of Harvey and Phillips (1979) for regression models with ARMA(p,q) disturbances is described. Their estimation method is based on the state space model representation of the regression model with ARMA(p,q) disturbances. They show how the exact likelihood function of the observations can be calculated by using the so called Kalman filter equations. For the specific MA(1) case the authors demonstrate by means of Monte Carlo experiments that an approximate variant of the method performs nearly as well as the exact method. Since for a general ARMA(p,q) process it is more difficult to compute directly the exact initial values required for starting the recursive Kalman equations, the authors propose to apply the approximate method for the general ARMA(p,q) process as well. However, to make sure that this is a viable approach one needs, in fact, the exact initial values for the general ARMA(p,q) case. Based on the results from Chapter 2 formulae are derived for the calculation of these exact initial values for the general ARMA(p,q) case.

Although in the literature many derivations of the Kalman equations can be found, Chapter 4 starts after a general introduction on multivariate analysis with an alternative derivation of the Kalman filter equations, wherein the geometric concepts described in Chapter 2, play an important role. The derivation is based on a certain regression equation which, to my best knowledge, is not mentioned elsewhere in the literature. This specific regression equation directly explains the structure of the Kalman gain formula. Duncan and Horn (1972), for instance, give also a derivation of the Kalman equations based on regression results and the generalized least squares principle. However, their derivation is not valid in all cases, especially not when the measurement noise has zero variance; see Harvey (1981). It is also argued that a number of standard derivations in systems theory cannot directly be applied to the classical regression model because, in general, in such a case the starting values are random rather than fixed. Furthermore, it is pointed out that the derivation given in Hannan and Deistler (1988) suffers from two minor errors which compensate each other.

In Chapter 5 the results derived by Wansbeek and Kapteyn (1985) are generalized. Wansbeek and Kapteyn derive the appropriate lower triangular transformation matrix for the linear regression model with missing observations and disturbances that follow an AR(1) process. Based on the geometric concepts and the methodology of Chapter 2 formulae are derived for the elements of the appropriate transformation matrix for the AR(2) case which formulae can easily be generalized for the AR(p) case, p>2. Furthermore it is shown how the Kalman filter equations can be used to estimate the model parameters when observations are missing and the disturbances follow an ARMA(p,q) process. Besides, attention is paid to the likelihood ratio test and the Lagrange multiplier test and their usefulness for discriminating between the several types of autocorrelation when observations are missing.

In Chapter 6 the geometric distributed lag model and its Koyck transform are studied. Under the assumption that the disturbances of the original distributed lag model are white noise it is demonstrated how the results from the inconsistent OLS estimation of the Koyck transformed model can directly be used to obtain consistent estimates of the original model. Furthermore, a 3-step estimator is described for the general dynamic linear model with ARMA(1,1) disturbances, which method can easily be generalized for ARMA(p,q) disturbances. Similar to the Gauss-Newton maximization approach the proposed estimation method is based on a Taylor series expansion with a special adjustment for the first few observations. This might be of interest for data from small samples. Finally, attention is paid to a geometric interpretation of the Cramér-Rao inequality and its relationship to the Pythagorean theorem.

In Chapter 7 the test strategy of Thursby (1981) is considered. The test strategy is intended to discriminate between misspecification of the regression model under study and several types of autocorrelation among the disturbances. Unfortunately, Porter and Kashyap (1984) provided Monte Carlo results that illustrate how disturbance autocorrelation can lead to a high number of false detections of misspecification because the RESET test statistic in Thursby's test strategy has a high rejection frequency when the disturbances are correlated. Furthermore, in Knottnerus (1985) it is shown that Thursby's test strategy is incapable of detecting certain types of autocorrelation. This is of interest especially for the case that the data have a seasonal pattern. Also it is demonstrated that a specific test of nonlinear restrictions on the parameters is inconsistent and superfluous since these restrictions happen to be satisfied always irrespective of the type of autocorrelation. Next, the test strategy of Godfrey (1987) is described which does not suffer from the problems mentioned above. Chapter 7 concludes with some comments on Godfrey's test strategy.

With the exception of Chapter 3 all chapters are more or less self-contained. Although the formula of the autocovariance matrix of an ARMA(p,q) process in Chapter 2 is referred to in other chapters, it is not necessary for a better understanding of these chapters to read Chapter 2 first. When cross-references to equations in other chapters are made we adopt the convention to use three numbers, the first one indicating the chapter while the last two refer to the equation number in that chapter.

A final remark regards the Pythagorean illustration in Appendix A of this chapter. Apart from the well-known Cauchy-Schwarz inequality the applications themselves are covered in the various chapters of this volume; see the subject index. They are combined here in one figure for technical reasons. The Cauchy-Schwarz inequality is included because it is often used in the literature for the proof of the Cramér-Rao inequality. For a further explanation of the illustration the reader is referred to the corresponding chapters.

Appendix 1.A Pythagorean illustration

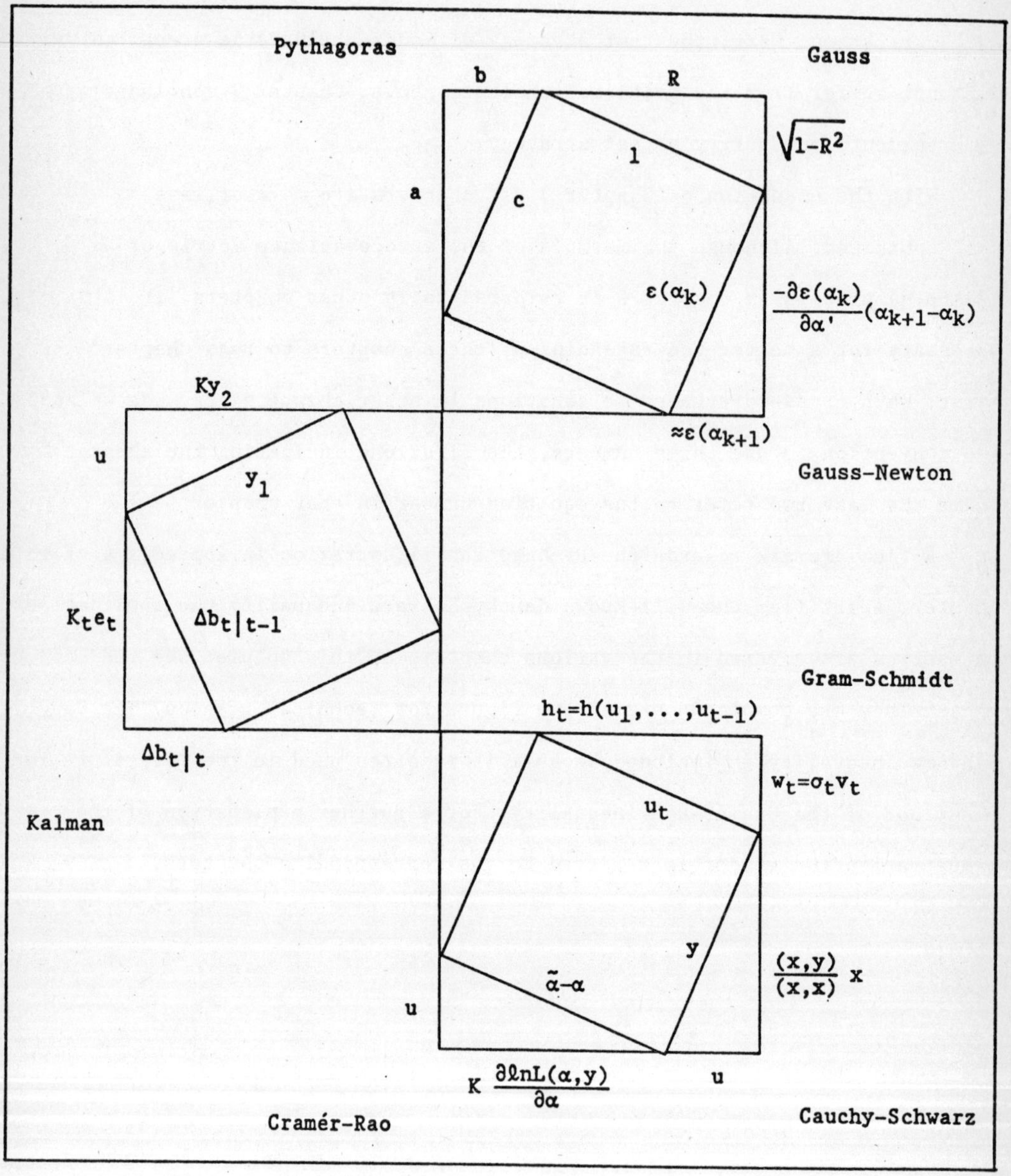

Figure 1. Illustration of the applications of the Pythagorean theorem

CHAPTER II

TRANSFORMATION MATRICES AND MAXIMUM LIKELIHOOD ESTIMATION OF REGRESSION MODELS WITH CORRELATED DISTURBANCES

2.1 Introduction

It often occurs that disturbances in a regression model are correlated. The context can be quite different. A first example is the system of n equations which are known as the seemingly unrelated regression equations, and where the disturbances of distinct equations are contemporaneously correlated. Another example is the simple regression model of a time series which consists of a single equation with correlated disturbances. In this study our main concern is the latter type of model.

One of the aims of this chapter is to find an appropriate transformation of the original regression model such that ordinary least squares (OLS) estimation of the transformed regression model yields the generalized least squares (GLS) estimators of the model parameters. In the case the disturbances follow an autoregressive (AR) process, expressions for an appropriate transformation are available. Also in case of a first order moving-average (MA) process Balestra (1980) and Pesaran (1973) give explicit but mutually dissimilar expressions for an appropriate transformation. However, in case the disturbances are generated by a higher order moving-average process or a mixed autoregressive moving-average (ARMA) process, an expression for an appropriate transformation matrix is not available with the exception of the ARMA(1,1) process; see Tiao and Ali (1971).

In this chapter an expression for the autocovariance matrix of disturbances which follow an ARMA(p,q) process is derived. Also expressions for the determinant of the autocovariance matrix and its inverse matrix are given where no matrices larger than q×q are inverted. Of the corresponding transformation matrix only the first p rows are to be determined by means of a recursive method. Hence, exact maximum likelihood (ML) estimation can be performed by applying a standard numerical maximization routine. Also it is shown that a simple recursive method, due to Levinson (1946) and Durbin (1960), can be used to invert the autocovariance matrix of an ARMA(p,q) process. Besides a more general recursive method is presented to invert an arbitrary covariance matrix. Many authors present methods for inverting covariance matrices, both exact and approximate, but not always they pay attention to the determinant which is indispensable for exact ML estimation; see Zinde-Walsh (1988). A list of references to methods for inverting covariance matrices in case of the ARMA model is given in Shaman (1975) and De Gooijer (1978).

Several authors have proposed methods for calculating GLS estimators for regression models with ARMA(p,q) disturbances. Amemiya (1973) gives a GLS estimator utilizing the estimated covariance matrix of the disturbances in a regression equation. Harvey and Phillips (1979) give a GLS estimator based on the Kalman filter algorithm which is relatively easy for MA(1) disturbances, but it can become tedious for higher-order ARMA disturbances. The method proposed here, is related to Ansley (1979). The main difference is that Ansley's method is based on an algorithm for calculating the Cholesky decomposition of the autocovariance matrix of ARMA(p,q) disturbances, whereas we straightforwardly derive formulae by employing the Gram-Schmidt orthogonalization.

The outline of this chapter is as follows. In Section 2 the algebraic problem of ML estimation is formulated. In Section 3 a dual problem in linear algebra is described. In Section 4 recursive formulae are derived for the inversion of the Cholesky roots of an arbitrary covariance matrix and a symmetric Toeplitz matrix, respectively. Furthermore, a recursive and numerically stable method is suggested for inverting any nonsingular matrix. Formulae for transforming the data of a regression model in such a way that OLS estimation of the transformed model produces the GLS estimators, are given in the Sections 5, 6, and 7, where the disturbances of the regression model follow an MA(1), MA(q) and ARMA(p,q) process, respectively. In Section 7 also an expression is derived for the autocovariance matrix of an arbitrary but stationary ARMA process. Besides, attention is paid to refinements for the case of certain types of seasonal processes.

2.2 The algebraic problem

The single equation linear regression model of time series may be written in matrix notation as

$$y = X\alpha + u \tag{2.1}$$

where y and u are n×1 column vectors, with n being the number of observations. X is an n×k matrix of regression variables and α is a k×1 parameter vector. The elements u_i of u are assumed to be normally distributed with $Eu_i=0$ and $Eu_iu_j=\sigma^2\gamma_{ij}$. If we assume covariance stationarity, then we have $Eu_iu_j=\sigma^2\gamma_{|i-j|}$ $(i,j=1,2,\ldots,n)$. The justification of the use of $\sigma^2\gamma_{ij}$ instead of γ_{ij} is that the log likelihood function reduces to a quadratic form after σ^2 has been concentrated out; see Section 2 in Chapter 4, viz. (4.2.16) and (4.2.17).

The system of n equations which are known as the seemingly unrelated regression equations (SURE) may be written in matrix notation as

$$\begin{bmatrix} y_1 \\ y_2 \\ \cdot \\ \cdot \\ \cdot \\ y_n \end{bmatrix} = \begin{bmatrix} X_1 & 0 & \dots & 0 \\ 0 & X_2 & \dots & 0 \\ \cdot & \cdot & & \cdot \\ \cdot & \cdot & & \cdot \\ \cdot & \cdot & & \cdot \\ 0 & 0 & & X_n \end{bmatrix} \begin{bmatrix} \alpha_1 \\ \alpha_2 \\ \cdot \\ \cdot \\ \cdot \\ \alpha_n \end{bmatrix} + \begin{bmatrix} u_1 \\ u_2 \\ \cdot \\ \cdot \\ \cdot \\ u_n \end{bmatrix} \tag{2.2}$$

where y_i and u_i are T×1 vectors, X_i is a matrix of order (T,k_i), and α_i is a k_i×1 vector (i=1,2,...,n). Contrary to the single equation model n is here the number of equations of the SURE model, whereas T denotes the number of observations of each separate equation. As regards the disturbances in (2.2) we assume the elements in u_i to obey the classical assumptions

$$Eu_i = 0 \tag{2.3a}$$

$$Eu_i u_j' = \sigma^2 \gamma_{ij} I_T, \quad i,j=1,2,\dots,n \tag{2.3b}$$

where I_T is the identity matrix of order T.

For the sake of uniform notation define for both model (2.1) and model (2.2) the n×n matrix $\sigma^2\Gamma$ as the covariance matrix with elements $\sigma^2\gamma_{ij}$. Then in the context of ML estimation of (2.1) and (2.2) the algebraic problem arises to construct a matrix P such that

$$P\Gamma P' = I_n \tag{2.4}$$

If Γ is nonsingular equation (2.4) is equivalent to

$$\Gamma^{-1} = P'P \tag{2.5}$$

It should be noted that P is not uniquely determined unless P is subject to certain restrictions. For instance, if we suppose P to be lower triangular then P is uniquely determined indeed and (2.5) is called the Cholesky decomposition of Γ^{-1}.

After finding a matrix P that satisfies (2.4), model (2.1) may be written as

$$Py = PX\alpha + Pu \tag{2.6}$$

and OLS estimation of (2.6) yields the ML estimator of α provided that Γ is known. To illustrate how the transformation must be performed in the case of model (2.2) we briefly write (2.2) similar to (2.1) as

$$y = X\alpha + u \tag{2.7}$$

with $Euu'=\sigma^2\Gamma\otimes I_T$. Now the appropriate transformation of (2.7) may be written as

$$(P\otimes I_T)y = (P\otimes I_T)X\alpha + (P\otimes I_T)u \tag{2.8}$$

where $\otimes$ stands for the Kronecker product. If we now define $v=(P\otimes I_T)u$, then it follows from (2.4) that $Evv'=\sigma^2 I_n\otimes I_T$. Hence, OLS estimation of (2.8) produces the ML estimators of the α_j in model (2.2).

It should be noted that OLS estimation of the transformed regression model produces the ML estimators of the regression coefficients, only if the parameters of the covariance matrix of the disturbances are known. If this is not the case then the Jacobian of the transformation has to be taken into account. This straightforwardly follows from maximizing the log likelihood function; see Ansley (1979) and Beach and MacKinnon (1978). If the parameters of the covariance matrix are unknown, then the adapted transformation matrix P_a is equal to $P_a=|P|^{-1/n}P$, where $|P|$ denotes the determinant of P. Therefore, it appears to be relevant to find an expression for the determinant of P as well.

In case of model (2.1) it can be shown that maximizing the log likelihood function is equivalent to minimizing SSR, where SSR is defined as $SSR=(P_a y-P_a X\alpha)'(P_a y-P_a X\alpha)$; see, for instance, Ansley (1979) as well as Section 2 of Chapter 4, especially (4.2.16). Concentrating out α yields $SSR=z'z$, where $z=(I_n-Q(Q'Q)^{-1}Q')w$ with $Q=P_a X$ and $w=P_a y$. This latter expression for SSR can be minimized with respect to the covariance matrix parameters by applying a standard nonlinear least squares algorithm. Similar results can be derived for the SURE model (2.2). By applying Laplace's expansion theorem we obtain $|P\otimes I_T|=|P|.|P\otimes I_{T-1}|=\ldots=|P|^T$. Hence, we have $|P\otimes I_T|^{-1/(nT)}(P\otimes I_T)=P_a\otimes I_T$ with $P_a=|P|^{-1/n}P$ and consequently, we obtain for the SURE model $SSR=z'z$ with $z=(I_{nT}-Q(Q'Q)^{-1}Q')w$, where $Q=(P_a\otimes I_T)X$ and $w=(P_a\otimes I_T)y$.

Remark. Note that according to Laplace's theorem we also have $|I_n\otimes B|=|B|^n$. Hence, more generally we obtain $|A\otimes B|=|(A\otimes I_m)(I_n\otimes B)|=|A|^m.|B|^n$, where A and B are square matrices of order n and m, respectively. Theil (1971, p. 303) derives the same result in a different way. He compares the eigenvalues of $A\otimes B$ with those of A and B.

2.3 A dual problem

The set of n normally distributed stochastic variables $u_1,u_2,\ldots,u_n$ with $Eu_t=0$ and $Eu_t u_j=\sigma^2\gamma_{tj}$ may be considered as a basis of a linear vector space S of dimension n over the real number field, provided that the n×n covariance matrix Γ with elements γ_{tj} is nonsingular; in the sequel of this chapter we set for simplicity the scalar factor $\sigma^2=1$. The inner product of two arbitrary elements v_1 and v_2 of S is defined as $(v_1,v_2)=Ev_1v_2$. Two elements v_1 and v_2 of S are said to be orthogonal if $(v_1,v_2)=0$. The norm or length of an element v_1 of S is defined as $\|v_1\|=(v_1,v_1)^{1/2}$, i.e., the standard deviation of v_1. Finally the Euclidean distance between two elements v_1 and v_2 of S is defined as

$d(v_1,v_2)=\|v_1-v_2\|$. Hence, the linear vector space S can be seen as a metric space as well. The introduction here of the abstract concept of a linear vector space is not strictly necessary, it only simplifies the mathematical presentation of the results because of the geometric interpretation of S, whereas, in fact, S is the set of all stochastic variables which can be written as a linear combination of $u_1,u_2,...,u_n$; see Appendix A for a formal definition of a linear vector space. Furthermore it should be noted that n normally distributed variables are stochastically independent if and only if their covariance matrix is a nonsingular diagonal matrix, while in a linear vector space n elements $v_1,v_2,...,v_n$ are said to be linearly independent if and only if $c_1v_1+...+c_nv_n=0$ implies that all c_t are zero (t=1,2,...,n). The latter definition of independence is more general than the former since linear independence only requires that the covariance matrix Evv' is nonsingular.

An alternative formulation of the problem in Section 2 of finding a matrix P such that $P\Gamma P'=I_n$ is given by the following dual problem. Find a subset of n elements of S, $v_1,v_2,...,v_n$ and a matrix P with elements p_{tj} which satisfy the following constraints

$$(v_t,v_j) = Ev_tv_j = 1, \qquad t=j \tag{3.1a}$$

$$(v_t,v_j) = Ev_tv_j = 0, \qquad t\neq j \tag{3.1b}$$

$$v_t = \sum_{j=1}^{n} p_{tj}u_j, \qquad t=1,2,...,n \tag{3.1c}$$

After finding these elements of S the result of the orthonormalization process may be summarized in matrix notation as $v=Pu$ and $Evv'=I_n$, where u and v are n×1 vectors with elements u_t and v_t, respectively.

The problem of finding a matrix P can be solved as follows. Define h_t as the projection of u_t onto the subspace S_{t-1} which is spanned by $u_1,u_2,...,u_{t-1}$ (t=2,3,...,n). Let w_t be defined as the corresponding perpendicular if $t\geqq 2$.

$$w_1 = u_1 \tag{3.2a}$$

$$w_t = u_t - h_t, \quad t \geqq 2 \tag{3.2b}$$

Then w_t and u_j are orthogonal if $j<t$. Furthermore w_t and w_j are orthogonal, because w_j is a linear combination of $u_1,u_2,\ldots,u_j$. Hence, $w_1,\ldots,w_n$ can be seen as the Gram-Schmidt orthogonalization of $u_1 \ldots,u_n$. Define

$$v_t = \frac{w_t}{\|w_t\|}, \quad t = 1,2,\ldots,n \tag{3.3}$$

then the system $v_1,v_2,\ldots,v_n$ is an orthonormal basis of the linear vector space S.

If we now write h_t as $h_t=\beta_{t1}u_{t-1}+\ldots+\beta_{t,t-1}u_1$, then in view of (3.2b) we have

$$u_t = \beta_{t1}u_{t-1} + \ldots + \beta_{t,t-1}u_1 + w_t \tag{3.4}$$

and the β_{tj} can be found by minimizing $\|w_t\|^2=Ew_t^2$ or, equivalently, by solving the corresponding normal equations $Ew_tu_{t-j}=0$ $(j=1,\ldots,t-1)$, yielding the so called least squares solution

$$\begin{bmatrix} \beta_{t1} \\ \cdot \\ \cdot \\ \cdot \\ \beta_{t,t-1} \end{bmatrix} = \begin{bmatrix} \gamma_{t-1,t-1} & \cdots & \gamma_{1,t-1} \\ \cdot & & \cdot \\ \cdot & & \cdot \\ \cdot & & \cdot \\ \gamma_{t-1,1} & \cdots & \gamma_{11} \end{bmatrix}^{-1} \begin{bmatrix} \gamma_{t,t-1} \\ \cdot \\ \cdot \\ \cdot \\ \gamma_{t1} \end{bmatrix} \tag{3.5}$$

As regards the length of w_t we have for $t>1$

$$\begin{aligned} \|w_t\|^2 &= Ew_t^2 \\ &= E(u_t - h_t)w_t \\ &= Eu_tw_t \\ &= Eu_t(u_t - \beta_{t1}u_{t-1} - \ldots - \beta_{t,t-1}u_1) \end{aligned}$$

$$= \gamma_{tt} - \sum_{j=1}^{t-1} \beta_{tj}\gamma_{t,t-j} \tag{3.6}$$

Define $\sigma_t = \|w_t\|$, then it follows from (3.2) and (3.6)

$$\sigma_1 = \gamma_{11}^{1/2} \tag{3.7a}$$

$$\sigma_t = (\gamma_{tt} - \sum_{j=1}^{t-1} \beta_{tj}\gamma_{t,t-j})^{1/2}, \quad t>1 \tag{3.7b}$$

It follows from (3.3) and (3.4) that

$$v_t = w_t/\sigma_t \tag{3.8a}$$

$$= (u_t - \beta_{t1}u_{t-1} - \dots - \beta_{t,t-1}u_1)/\sigma_t \tag{3.8b}$$

Hence, the lower triangular transformation matrix P with elements p_{tj} can be written as

$$P = \begin{bmatrix} \frac{1}{\sigma_1} & 0 & 0 \ \dots & 0 \\ -\frac{\beta_{21}}{\sigma_2} & \frac{1}{\sigma_2} & 0 \ \dots & 0 \\ -\frac{\beta_{32}}{\sigma_3} & -\frac{\beta_{31}}{\sigma_3} & \frac{1}{\sigma_3} \ \dots & 0 \\ \dots & & & \\ -\frac{\beta_{n,n-1}}{\sigma_n} & \dots & -\frac{\beta_{n1}}{\sigma_n} & \frac{1}{\sigma_n} \end{bmatrix} \tag{3.9}$$

From (3.7) and (3.9) it can be seen that P is determined by the β_{tj} and (3.9) can be expressed as

$$p_{tj} = -\frac{\beta_{t,t-j}}{\sigma_t}, \quad 1 \leqq t,j \leqq n \tag{3.10}$$

where $\beta_{t0}=-1$ and $\beta_{t,t-j}=0$ if $j>t$. Finally it follows from (3.9) that the determinant of P is given by

$$|P| = \prod_{t=1}^{n} \sigma_t^{-1} \tag{3.11}$$

Since $\Gamma^{-1}=P'P$ we obtain

$$|\Gamma| = \prod_{t=1}^{n} \sigma_t^{2} \tag{3.12}$$

2.4 Recursive methods for calculating the transformation matrix P

2.4.1 A recursive algorithm for calculating P

We assume the γ_{tj} to be known and note once more that we require P to be a lower triangular matrix. Because of (3.7a) we set

$$p_{11} = \gamma_{11}^{-1/2} \tag{4.1}$$

This is the initial stage of the algorithm. Next we describe the recursion.

Assuming that we already have an orthonormal basis $v_1,v_2,\dots,v_t$ of the subspace S_t that is spanned by $u_1,u_2,\dots,u_t$ and that v_k can be written as

$$v_k = \sum_{j=1}^{k} p_{kj}u_j, \quad 1 \leqq k \leqq t \tag{4.2}$$

we can calculate the elements $p_{t+1,j}$ of the $(t+1)^{th}$ row of the nxn matrix P as follows. We first write the projection h_{t+1} of u_{t+1} onto the subspace S_t as

$$h_{t+1} = \sum_{k=1}^{t} d_{t+1,k}v_k \tag{4.3}$$

and define w_{t+1} according to (3.2) as $w_{t+1}=u_{t+1}-h_{t+1}$. Since $v_1,\ldots,v_t$ are an orthonormal basis of S_t, we have

$$d_{t+1,k} = Eu_{t+1}v_k \tag{4.4}$$

Substitution of (4.2) into (4.4) gives

$$d_{t+1,k} = \sum_{j=1}^{k} p_{kj}\gamma_{t+1,j} \tag{4.5}$$

Clearly, we have according to the Pythagorean theorem

$$\|w_{t+1}\|^2 + \|h_{t+1}\|^2 = \|u_{t+1}\|^2 \tag{4.6}$$

as $E\{w_{t+1}h_{t+1}\}=0$. Since $v_1,v_2,\ldots,v_t$ are orthonormal, it follows from (4.3) and (4.6) that σ_{t+1}, the length of w_{t+1}, equals

$$\sigma_{t+1} = (\gamma_{t+1,t+1} - \sum_{k=1}^{t} d^2_{t+1,k})^{½} \tag{4.7}$$

If we define

$$v_{t+1} = \frac{w_{t+1}}{\sigma_{t+1}} \tag{4.8}$$

then $v_1,v_2,\ldots,v_{t+1}$ can be considered as an orthonormal basis of the subspace S_{t+1}, spanned by $u_1,\ldots,u_{t+1}$. From (3.2b), (4.2), (4.3) and (4.8) it follows that v_{t+1} can be written as

$$\begin{aligned} v_{t+1} &= \frac{1}{\sigma_{t+1}} (u_{t+1} - \sum_{k=1}^{t} d_{t+1,k}v_k) \\ &= \frac{1}{\sigma_{t+1}} (u_{t+1} - \sum_{k=1}^{t} \sum_{j=1}^{k} d_{t+1,k}p_{kj}u_j) \end{aligned}$$

$$= \frac{1}{\sigma_{t+1}} \{u_{t+1} - \sum_{j=1}^{t} (\sum_{k=j}^{t} d_{t+1,k}p_{kj})u_j\} \tag{4.9}$$

From (4.9) it follows that

$$p_{t+1,j} = \frac{1}{\sigma_{t+1}}, \quad \text{if } j=t+1 \tag{4.10a}$$

$$p_{t+1,j} = -\frac{1}{\sigma_{t+1}} \sum_{k=j}^{t} d_{t+1,k}p_{kj}, \quad \text{if } j<t+1 \tag{4.10b}$$

where $d_{t+1,k}$ and σ_{t+1} are defined in (4.5) and (4.7), respectively.

From (4.1) and (4.10) a recursive method follows for the calculation of matrix P. However, it should be noted that the numerical stability might be poor, if Γ is nearly singular and n is large; see Stoer (1976, pp. 169-171). Notwithstanding, these formulae seem to be practical in the context of a set of seemingly unrelated regression equations, if the number of equations is small. In Section 7 they will also appear to be useful for deriving the first p rows of the transformation matrix P, if the disturbances follow an ARMA(p,q) process. In case of numerical instability it may be useful to perform a second step which consists of the calculation of a second matrix, say T, in the same manner as before by means of (4.1), (4.5), (4.7) and (4.10). One only has to replace the elements γ_{tj} of Γ by the elements of $\Gamma^*=P\Gamma P'$. Then we obtain $TP\Gamma P'T'=I_n$ and it is hoped that a more appropriate transformation matrix is given by TP instead of P.

Further, it should be noted that application of this recursive method yields the inverse of P as well. If we define $L=P^{-1}$ the transformation can be written in matrix notation as v=Pu, and hence, u=Lv. Then from the first equality of (4.9) it can be seen that L equals

$$L = \begin{bmatrix} \sigma_1 & 0 & 0 & \cdots & 0 \\ d_{21} & \sigma_2 & 0 & \cdots & 0 \\ d_{31} & d_{32} & \sigma_3 & \cdots & 0 \\ \cdots & & & & \\ d_{n1} & d_{n2} & \cdots & d_{n,n-1} & \sigma_n \end{bmatrix}$$

In terms of the Cholesky decomposition we have $\Gamma = LL'$.

2.4.2 The recursive Levinson-Durbin algorithm

Next, we assume Γ to be a symmetric Toeplitz matrix, i.e., $\gamma_{kj} = \gamma_{|k-j|}$. Then (3.5) may be rewritten as

$$\begin{bmatrix} \beta_{t1} \\ \beta_{t2} \\ \cdot \\ \cdot \\ \cdot \\ \beta_{t,t-1} \end{bmatrix} = \begin{bmatrix} \gamma_0 & \gamma_1 & \cdots & \gamma_{t-2} \\ \gamma_1 & \gamma_0 & \cdots & \gamma_{t-3} \\ \cdot & \cdot & & \cdot \\ \cdot & \cdot & & \cdot \\ \cdot & \cdot & & \cdot \\ \gamma_{t-2} & \gamma_{t-3} & \cdots & \gamma_0 \end{bmatrix}^{-1} \begin{bmatrix} \gamma_1 \\ \gamma_2 \\ \cdot \\ \cdot \\ \cdot \\ \gamma_{t-1} \end{bmatrix} \tag{4.11}$$

and (3.7) can be written as

$$\sigma_1 = \gamma_0^{1/2} \tag{4.12a}$$

$$\sigma_t = \left(\gamma_0 - \sum_{j=1}^{t-1} \beta_{tj}\gamma_j\right)^{1/2}, \quad 2 \leq t \leq n \tag{4.12b}$$

The formulae (4.11) and (4.12) can be used to calculate an appropriate transformation matrix in case the disturbances follow an arbitrary but stationary ARMA(p,q) process. Although the context of (4.11) is rather different, its structure is exactly the structure of the well-known Yule-Walker equations. Levinson (1946) and Durbin (1960) give a recursive algorithm for calculating the p autoregressive parameters of an AR(p) process. However, this simple algorithm can be used here to calculate the β_{tj} if there is an ARMA(p,q) process among the disturbances of the regression model under consideration. As a matter of fact we first have $\beta_{21} = \gamma_1/\gamma_0$. Then applying the recursive Levinson-Durbin (LD) algorithm we obtain

$$\begin{bmatrix}\beta_{t+1,1}\\ \beta_{t+1,2}\\ \cdot\\ \cdot\\ \cdot\\ \beta_{t+1,t-1}\end{bmatrix} = \begin{bmatrix}\beta_{t1}\\ \beta_{t2}\\ \cdot\\ \cdot\\ \cdot\\ \beta_{t,t-1}\end{bmatrix} - \beta_{t+1,t}\begin{bmatrix}\beta_{t,t-1}\\ \beta_{t,t-2}\\ \cdot\\ \cdot\\ \cdot\\ \beta_{t1}\end{bmatrix}, \quad t=2,3,... \tag{4.13a}$$

where

$$\beta_{t+1,t} = (\gamma_t - \sum_{j=1}^{t-1} \beta_{tj}\gamma_{t-j})/\sigma_t^2 \tag{4.13b}$$

After determining the σ_t and β_{tj} from (4.12) and (4.13), respectively, the matrix P can be calculated from (3.9). See Section 7 for an analytical expression for the autocovariances of a general ARMA model.

2.4.3 A supplementary Levinson-Durbin algorithm

Box and Jenkins (1976, p. 65) remark that the LD algorithm becomes very sensitive to rounding errors and should not be used if the values of the parameters are close to the nonstationarity boundary. However, the authors do not explain what criteria can be used to detect this numerical instability. The following device might shed some light on this. After β_{tj} and σ_t have been calculated the lower triangular matrix P can be determined according to (3.9). Since $P\Gamma P'=I_n$ it can easily be checked whether inaccuracies have been introduced in the calculations of the β_{tj} by comparing I_n and $\bar{P}\Gamma\bar{P}'$, with $\bar{P}$ being the numerical result of the LD algorithm. If inaccuracies are found it is worthwhile to calculate in a second step the matrix T with the property $T(\bar{P}\Gamma\bar{P}')T'=I_n$; see also the discussion below (4.10). Defining $P^*=T\bar{P}$ it can easily be seen from (3.9) that new values of the β_{tj} can be obtained according to $\beta_{tj}=-p^*_{t,t-j}/p^*_{tt}$, j=1,2,...,t-1.

2 4.4 Inversion of an arbitrary nonsingular matrix

In this subsection we shall briefly pay some attention to the inversion of an arbitrary nonsingular matrix U and the calculation of its determinant. For that purpose we first write

$$U^{-1} = \Gamma^{-1}U' \tag{4.14}$$

where $\Gamma = U'U$. Then employing (4.10) we obtain a matrix P such that upon substitution of (2.5) into (4.14) the latter formula can be written as

$$U^{-1} = P'P\,U' \tag{4.15a}$$

$$= P'T'TPU' \text{ (in the case of numerical instability)} \tag{4.15b}$$

Furthermore, it follows from (3.11) and (4.15a) that the absolute value of the determinant of U is equal to

$$\text{abs}(|U|) = |P^{-1}| = \prod_{t=1}^{n} \sigma_t \tag{4.16}$$

Since we have $\sigma_\iota = \|u_\iota - h_\iota\|$, i.e., the length of the corresponding perpendicular, (4.16) reflects the well-known geometric interpretation of the determinant as the volume of the n-dimensional hyperparallelepiped $G=\{\sum_{t=1}^{n} \lambda_t u_t ; 0<\lambda_t<1\}$, where u_t is the tth column vector of matrix U. If $V(u_1,\ldots,u_n)$ is defined as the volume of the hyperparallelepiped G, then (4.16) may be rewritten alternatively

$$|\Gamma| = \{V(u_1,u_2,\ldots,u_n)\}^{-1} \tag{4.17}$$

This expression will appear to be meaningful in Section 7 where we will use it in the context of the linear vector space S, introduced in Section 3. Finally, it should be noted that it can be seen from (2.5) that $|\Gamma|$ equals the squared volume of this hyperparallelepiped

$$|\Gamma| = \{V(u_1,\ldots,u_n)\}^2 \tag{4.18}$$

2.5 The matrix P in the case of MA(1) disturbances

2.5.1 The matrix P

Henceforth we only consider the simple time series regression model

$$y_t = x_t'\alpha + u_t, \quad t=1,2,\ldots,n \tag{5.1}$$

where x_t and α are k×1 vectors, and the u_t follow a stationary ARMA(p,q) process.

In this section we shall derive the matrix P for the case where the u_t in (5.1) follow an MA(1) process

$$u_t = \varepsilon_t - \lambda\varepsilon_{t-1} \tag{5.2}$$

Without loss of generality we assume that the ε_t are independent and identically distributed (iid) N(0,1). The autocovariances follow from (5.2)

$$\gamma_0 = 1+\lambda^2 \tag{5.3a}$$

$$\gamma_1 = -\lambda \tag{5.3b}$$

$$\gamma_k = 0\ , \quad k\geq 2 \tag{5.3c}$$

Substitution of (5.3) into (4.11) gives

$$\begin{bmatrix} \beta_{t1} \\ \beta_{t2} \\ \cdot \\ \cdot \\ \cdot \\ \\ \beta_{t,t-1} \end{bmatrix} = \begin{bmatrix} 1+\lambda^2 & -\lambda & 0 & \cdots & 0 & 0 \\ -\lambda & 1+\lambda^2 & -\lambda & \cdots & 0 & 0 \\ 0 & -\lambda & 1+\lambda^2 & \cdots & 0 & 0 \\ \cdot & \cdot & \cdot & & \cdot & \cdot \\ \cdot & \cdot & \cdot & & \cdot & \cdot \\ 0 & 0 & 0 & \cdots & 1+\lambda^2 & -\lambda \\ 0 & 0 & 0 & \cdots & -\lambda & 1+\lambda^2 \end{bmatrix}^{-1} \begin{bmatrix} -\lambda \\ 0 \\ \cdot \\ \cdot \\ \cdot \\ \cdot \\ 0 \end{bmatrix} \tag{5.4}$$

Denote the txt autocovariance matrix of an MA(1) process as Γ_t and define $d_t=|\Gamma_t|$, then by expansion of the determinant of Γ_t with respect to the elements of the first column, the following recursive formula can be established

$$d_t = (1+\lambda^2)d_{t-1} - \lambda^2 d_{t-2}, \quad t>1 \tag{5.5}$$

with $d_0=1$. Now by applying the principle of mathematical induction we obtain the well-known formula for the determinant of Γ_t

$$\begin{aligned} d_t &= 1 + \lambda^2 + \ldots + \lambda^{2t} \\ &= \frac{1-\lambda^{2t+2}}{1-\lambda^2}, \quad t \geqq 1 \end{aligned} \tag{5.6}$$

Further, it can be seen from (5.4) that the cofactor of the jth element of the first column (row) of Γ_{t-1} is equal to

$$cf(j,1) = \lambda^{j-1}\, d_{t-j-1}, \quad 1 \leqq j \leqq t-1 \tag{5.7}$$

From (5.4) and (5.7) it follows

$$\beta_{tj} = -\lambda^j \frac{d_{t-j-1}}{d_{t-1}}, \quad 1 \leqq j \leqq t-1 \tag{5.8}$$

The last step in the derivation of a matrix P according to (3.9) is the calculation of the σ_t. Substitution of (5.3) and (5.8) into (4.12) yields

$$\sigma_1^2 = 1 + \lambda^2 \tag{5.9a}$$

$$\sigma_t^2 = 1 + \lambda^2 - \lambda^2 \frac{d_{t-2}}{d_{t-1}}, \qquad t \geqq 2 \tag{5.9b}$$

From (5.5) and (5.9) it can be seen that

$$\sigma_t^2 = \frac{d_t}{d_{t-1}}, \qquad 1 \leqq t \leqq n \tag{5.10}$$

Now it follows from (3.10), (5.8) and (5.10) that for MA(1) disturbances the transformation matrix P may be written as

$$P = \begin{bmatrix} \frac{d_0}{(d_1 d_0)^{1/2}} & & \cdots & 0 \\ \frac{\lambda\, d_0}{(d_2 d_1)^{1/2}} & \frac{d_1}{(d_2 d_1)^{1/2}} & \cdots & 0 \\ \cdot & \cdot & & \cdot \\ \cdot & \cdot & & \cdot \\ \cdot & \cdot & & \cdot \\ \frac{\lambda^{n-1} d_0}{(d_n d_{n-1})^{1/2}} & \frac{\lambda^{n-2} d_1}{(d_n d_{n-1})^{1/2}} & \cdots & \frac{d_{n-1}}{(d_n d_{n-1})^{1/2}} \end{bmatrix} \tag{5.11}$$

Balestra (1980) gives the same matrix P but without a derivation. In fact he verifies the identity $P\Gamma_n P' = I_n$.

2.5.2 A new derivation of the inverse of the autocovariance matrix of an MA(1) process

By employing (5.11) a simple expression for Γ^{-1} can be derived. Define $M=\Gamma^{-1}$, then as a matter of fact we have $m_{11}=d_{n-1}/d_n$ and as $m_{11}=(P'P)_{11}$, we obtain $(P'P)_{11}=d_{n-1}/d_n$. By substituting (5.11) into the latter equation the following formula can be established

$$\sum_{k=1}^{n} \frac{\lambda^{2k-2}}{d_k d_{k-1}} = \frac{d_{n-1}}{d_n} \tag{5.12}$$

Now it follows from (5.11) and (5.12) that m_{ij} can be expressed as

$$\begin{aligned} m_{ij} &= \sum_{k=j}^{n} p_{ki}p_{kj} \\ &= \frac{\lambda^2 d_{i-1}d_{j-1}}{\lambda^{i+j}} \times \sum_{k=j}^{n} \frac{\lambda^{2k-2}}{d_k d_{k-1}} \\ &= \frac{\lambda^2}{\lambda^{i+j}} d_{i-1}d_{j-1}\left(\frac{d_{n-1}}{d_n} - \frac{d_{j-2}}{d_{j-1}}\right), \quad 1 \leqq i \leqq j \leqq n \end{aligned} \tag{5.13}$$

where use has been made of $p_{ki}=\lambda^{k-i}d_{i-1}/(d_k d_{k-1})^{½}$, $k \geqq i$. Substitution of (5.6) into (5.13) yields

$$m_{ij} = \frac{(1-\lambda^{2i})(\lambda^{2j} - \lambda^{2n+2})}{(1-\lambda^2)(1-\lambda^{2n+2})\lambda^{i+j}}, \quad 1 \leqq i \leqq j \leqq n \tag{5.14}$$

As M is symmetric, we have $m_{ji}=m_{ij}$. Arato (1961) seems to be the first author who gave an analytical expression for Γ^{-1}. Later it was rediscovered by Balestra in 1972; see Balestra (1980, p. 382) and De Gooijer (1984, p. 198).

When employing the algorithm proposed by Ansley (1979) for calculating the log likelihood function in case of an MA(1) process, the Cholesky root L of the covariance matrix of such a process has to be calculated by means of a specific algorithm. However, by definition we have $L = P^{-1}$ and the inverse of

P straightforwardly follows from (5.11). Hence, we obtain

$$L = \begin{bmatrix} \sigma_1 & 0 & 0 & \cdots & 0 & 0 \\ -\lambda/\sigma_1 & \sigma_2 & 0 & \cdots & 0 & 0 \\ 0 & -\lambda/\sigma_2 & \sigma_3 & \cdots & 0 & 0 \\ \cdot & \cdot & \cdot & & \cdot & \cdot \\ \cdot & \cdot & \cdot & & \cdot & \cdot \\ \cdot & \cdot & \cdot & & \cdot & \cdot \\ 0 & 0 & 0 & \cdots & \sigma_{n-1} & 0 \\ 0 & 0 & 0 & \cdots & -\lambda/\sigma_{n-1} & \sigma_n \end{bmatrix} \tag{5.15}$$

where σ_t is defined by (5.10) (t=1,2,...,n). Furthermore, since Lv=u we obtain $v_1=u_1/\ell_{11}$ and $v_t=(u_t-\ell_{t,t-1}v_{t-1})/\ell_{tt}$, t>1. Therefore, in order to apply the OLS estimation procedure the data are to be transformed according to

$$\bar{y}_1 = y_1/\ell_{11} \quad \text{and} \quad \bar{y}_t = (y_t-\ell_{t,t-1}\bar{y}_{t-1})/\ell_{tt}, \; t>1 \tag{5.16a}$$

$$\bar{x}_1 = x_1/\ell_{11} \quad \text{and} \quad \bar{x}_t = (x_t-\ell_{t,t-1}\bar{x}_{t-1})/\ell_{tt}, \; t>1 \tag{5.16b}$$

This transformation is also given by Balestra (1980).

2.6 The matrix P in the case of MA(q) disturbances

We first consider the case that the disturbances follow an MA(2) process

$$u_t = \varepsilon_t - \lambda_1\varepsilon_{t-1} - \lambda_2\varepsilon_{t-2} \tag{6.1}$$

The autocovariances γ_k are given by

$$\gamma_0 = 1 + \lambda_1^2 + \lambda_2^2 \tag{6.2a}$$

$$\gamma_1 = -\lambda_1(1-\lambda_2) \tag{6.2b}$$

$$\gamma_2 = -\lambda_2 \tag{6.2c}$$

$$\gamma_k = 0, \quad k>2 \tag{6.2d}$$

We will now rewrite (4.11) by premultiplying both sides by the $(t-1)\times(t-1)$ autocovariance matrix of an MA(2) process, using $\gamma_k=0$ for $k>2$. We obtain the following t-1 normal equations

$$\gamma_0\beta_{t1}+\gamma_1\beta_{t2}+\gamma_2\beta_{t3}=\gamma_1 \tag{6.3a}$$

$$\gamma_1\beta_{t1}+\gamma_0\beta_{t2}+\gamma_1\beta_{t3}+\gamma_2\beta_{t4}=\gamma_2 \tag{6.3b}$$

$$\gamma_2\beta_{tj}+\gamma_1\beta_{t,j+1}+\gamma_0\beta_{t,j+2}+\gamma_1\beta_{t,j+3}+\gamma_2\beta_{t,j+4}=0, \quad 1\leqq j\leqq t-5 \tag{6.3c}$$

$$\gamma_2\beta_{t,t-4}+\gamma_1\beta_{t,t-3}+\gamma_0\beta_{t,t-2}+\gamma_1\beta_{t,t-1}=0 \tag{6.3d}$$

$$\gamma_2\beta_{t,t-3}+\gamma_1\beta_{t,t-2}+\gamma_0\beta_{t,t-1}=0 \tag{6.3e}$$

The third equation is a homogeneous difference equation. A similar difference equation is given by Mentz (1976) and Nerlove et al. (1979, pp. 416-421) where it is used to invert more general band matrices of Toeplitz type. The solution of (6.3c) follows from the roots of the characteristic polynomial

$$0 = \gamma_2 + \gamma_1 x + \gamma_0 x^2 + \gamma_1 x^3 + \gamma_2 x^4 \tag{6.4}$$

In the specific case of MA(2) disturbances we obtain upon substitution of (6.2) into (6.4)

$$\begin{aligned} 0 &= -\lambda_2 - \lambda_1(1-\lambda_2)x + (1+\lambda_1^2+\lambda_2^2)x^2 - \lambda_1(1-\lambda_2)x^3 - \lambda_2 x^4 \\ &= (x^2-\lambda_1 x-\lambda_2)(1-\lambda_1 x-\lambda_2 x^2) \end{aligned} \tag{6.5}$$

The roots of (6.5) can be expressed as x_1, x_2, $1/x_1$, and $1/x_2$, respectively. Hence, the general solution of the difference equation is

$$\beta_{tj} = c_{t1}\psi_{1j} + c_{t2}\psi_{2j} + c_{t3}\psi_{3j} + c_{t4}\psi_{4j} \quad (t=5,6,\dots,n;\ j=1,2,\dots,t-1) \tag{6.6}$$

where

$\psi_{1j}=x_1^j$, $\psi_{2j}=x_2^j$, $\psi_{3j}=x_1^{-j}$, and $\psi_{4j}=x_2^{-j}$,

if $\lambda_1^2+4\lambda_2>0$ and hence $x_{12}=\frac{1}{2}\lambda_1\pm\frac{1}{2}(\lambda_1^2+4\lambda_2)^{\frac{1}{2}}$;

$\psi_{1j} = x^j$, $\psi_{2j} = jx^j$, $\psi_{3j} = x^{-j}$, and $\psi_{4j} = jx^{-j}$,

if $\lambda_1^2+4\lambda_2=0$ and hence $x=\lambda_1/2$;

$\psi_{1j} = x^j\sin(j\phi)$, $\psi_{2j} = x^j\cos(j\phi)$, $\psi_{3j} = x^{-j}\sin(j\phi)$, and

$\psi_{4j}=x^{-j}\cos(j\phi)$, if $\lambda_1^2+4\lambda_2<0$ and hence $x_{12}=x(\cos\phi \pm i\sin\phi)$ (6.7)

The four constants $c_{t1},\dots,c_{t4}$ in (6.6) can be solved from the four boundary conditions, which consist of the first two and the last two equations of (6.3). However, this direct approach is rather tedious; see Mentz (1976, pp. 434-436) and Nerlove et al. (1979). Therefore, we propose the following method which is more convenient, especially in the case at hand. Let β_{tj} also be defined as in (6.6) for $j\leqq 0$ and $j\geqq t$. Then because of the third equation of (6.3) the four boundary conditions of (6.3) can be written in an alternative manner as

$$\gamma_2\beta_{t,-1} + \gamma_1\beta_{t0} = -\gamma_1 \tag{6.8a}$$

$$\gamma_2\beta_{t0} = -\gamma_2 \tag{6.8b}$$

$$\gamma_2\beta_{tt} = 0 \tag{6.8c}$$

$$\gamma_1\beta_{tt} + \gamma_2\beta_{t,t+1} = 0 \tag{6.8d}$$

From (6.8c and d) it follows that $\beta_{tt}=\beta_{t,t+1}=0$. Therefore (6.8) may be rewritten in matrix notation as

$$\begin{bmatrix} \gamma_2 & \gamma_1 & 0 & 0 \\ 0 & \gamma_2 & 0 & 0 \\ 0 & 0 & 1 & 0 \\ 0 & 0 & 0 & 1 \end{bmatrix} \begin{bmatrix} \beta_{t,-1} \\ \beta_{t0} \\ \beta_{tt} \\ \beta_{t,t+1} \end{bmatrix} = \begin{bmatrix} -\gamma_1 \\ -\gamma_2 \\ 0 \\ 0 \end{bmatrix} \tag{6.9}$$

As

$$\begin{bmatrix} \gamma_2 & \gamma_1 \\ 0 & \gamma_2 \end{bmatrix}^{-1} \begin{bmatrix} \gamma_1 \\ \gamma_2 \end{bmatrix} = \begin{bmatrix} 0 \\ 1 \end{bmatrix}$$

we obtain $\beta_{t,-1}=0$, $\beta_{t0}=-1$ and $\beta_{tt}=\beta_{t,t+1}=0$ and hence, upon substitution of (6.6) into (6.9) the four constants become

$$\begin{bmatrix} c_{t1} \\ c_{t2} \\ c_{t3} \\ c_{t4} \end{bmatrix} = \begin{bmatrix} \psi_{1,-1} & \psi_{2,-1} & \psi_{3,-1} & \psi_{4,-1} \\ \psi_{10} & \psi_{20} & \psi_{30} & \psi_{40} \\ \psi_{1t} & \psi_{2t} & \psi_{3t} & \psi_{4t} \\ \psi_{1,t+1} & \psi_{2,t+1} & \psi_{3,t+1} & \psi_{4,t+1} \end{bmatrix}^{-1} \begin{bmatrix} 0 \\ -1 \\ 0 \\ 0 \end{bmatrix} \tag{6.10}$$

Now a nontrivial question naturally arises, i.e., the question of the validity of (6.6) if t=2,3,4. The answer is in the affirmative and this will be proved in Appendix B in the more general context of the MA(q) process.

Next we assume the u_t from (5.1) to follow an MA(q) process

$$u_t = \varepsilon_t - \lambda_1\varepsilon_{t-1} - \dots - \lambda_q\varepsilon_{t-q} \tag{6.11}$$

where ε_t is (iid) N(0,1). Define $\lambda_0=-1$, then we have

$$\gamma_k = \sum_{j=0}^{q-k} \lambda_j\lambda_{j+k}, \quad |k|\leq q \tag{6.12a}$$

$$\gamma_k = 0, \quad |k|>q \tag{6.12b}$$

The generalization of (6.6) is straightforward

$$\beta_{tj} = \sum_{p=1}^{2q} c_{tp}\psi_{pj}, \quad t>2q \quad \text{and} \quad j=1,2,\ldots,t-1 \tag{6.13}$$

Similar to (6.5) and (6.7) the ψ_{pj} follow from the solution of

$$x^q - \lambda_1 x^{q-1} - \ldots - \lambda_q = 0 \tag{6.14}$$

To illustrate how the ψ_{pj} are related to the roots of (6.14) we only consider the first root, say x_1, of (6.14). Regarding x_1 four possibilities have to be distinguished which follow from the theory of difference equations.

1. If x_1 is unique and real, then we have

$$\psi_{1j} = x_1^j \text{ and } \psi_{q+1,j} = x_1^{-j}$$

2. If x_1 is a real valued root of multiplicity m, i.e., $x_1=\ldots=x_m$ then we have

$$\psi_{pj} = j^{p-1}\, x_1^j, \quad 1\leqq p\leqq m$$
$$\psi_{pj} = j^{p-q-1}\, x_1^{-j}, \quad q+1\leqq p\leqq q+m$$

3. If x_1 is unique and complex, then also $\bar{x}_1$ is a root of (6.14) and x_1 can be written as $x_1=x(\cos\varphi + i \sin\varphi)$. By De Moivre's theorem we have

$$\psi_{1j} = x^j \sin(j\varphi), \quad \psi_{2j} = x^j \cos(j\varphi),$$
$$\psi_{q+1,j} = x^{-j} \sin(j\varphi), \quad \psi_{q+2,j} = x^{-j} \cos(j\varphi)$$

4. If x_1 is a complex valued root of multiplicity m then $\bar{x}_1$ is also a complex valued root of multiplicity m. And once more we set

$x_1=x(\cos\varphi + i \sin\varphi)$, then we have

$$\psi_{pj} = j^{p-1}\, x^j \sin(j\varphi), \quad 1\leqq p\leqq m$$
$$\psi_{pj} = j^{p-m-1}\, x^j \cos(j\varphi), \quad m+1\leqq p\leqq 2m$$
$$\psi_{pj} = x^{-2j}\, \psi_{p-q,j}\,, \quad q+1\leqq p\leqq q+2m$$

Finally, the constants in (6.13) follow from the generalization of (6.10)

$$\begin{bmatrix} c_{t1} \\ \cdot \\ \\ \cdot \\ \\ \cdot \\ c_{t,2q} \end{bmatrix} = - \begin{bmatrix} \psi_{1,-q+1} & \cdots & \psi_{2q,-q+1} \\ \cdots & & \\ \psi_{10} & & \psi_{2q,0} \\ \psi_{1t} & & \psi_{2q,t} \\ \cdots & & \\ \psi_{1,t+q-1} & \cdots & \psi_{2q,t+q-1} \end{bmatrix}^{-1} \begin{bmatrix} 0 \\ \vdots \\ 0 \\ 1 \\ 0 \\ \vdots \\ 0 \end{bmatrix} \tag{6.15}$$

Define the 2q×1 vector $c_t=(c_{t1},\ldots,c_{t,2q})'$ then assuming nonsingularity we can write (6.15) in partitioned matrix notation

$$c_t = - \begin{bmatrix} A_{11} & A_{12} \\ A_{21} & A_{22} \end{bmatrix}^{-1} \begin{bmatrix} e_q \\ 0 \end{bmatrix} \tag{6.16}$$

where e_q denotes the q×1 vector $(0,\ldots,0,1)'$. Using the well-known expression for the inverse of a partitioned matrix (6.16) can be written as

$$c_t = \begin{bmatrix} -I_q \\ A_{22}^{-1}A_{21} \end{bmatrix} \left[A_{11} - A_{12}A_{22}^{-1}A_{21}\right]^{-1} e_q \tag{6.17}$$

In order to obtain a solution for the β_{tj} (6.13) and (6.17) have to be combined for successively $j=1,2,\ldots,t-1$. This yields in matrix notation

$$\begin{bmatrix} \beta_{t1} \\ \beta_{t2} \\ \cdot \\ \cdot \\ \cdot \\ \beta_{t,t-1} \end{bmatrix} = \begin{bmatrix} \psi_{11} & \cdots & \psi_{2q,1} \\ \psi_{12} & \cdots & \psi_{2q,2} \\ \cdot & & \cdot \\ \cdot & & \cdot \\ \cdot & & \cdot \\ \psi_{1,t-1} & & \psi_{2q,t-1} \end{bmatrix} \begin{bmatrix} -I_q \\ A_{22}^{-1}A_{21} \end{bmatrix} \left[A_{11}-A_{12}A_{22}^{-1}A_{21}\right]^{-1} e_q \qquad (6.18)$$

With the aid of (6.18), (4.12) and (3.9) an appropriate transformation matrix P can now be determined. The Jacobian follows from (3.11) and (4.12).

The last step to be proved is the validity of (6.18) if $2 \leqq t \leqq 2q$. This question arises because the homogeneous difference equation and the 2q boundary conditions which follow from the generalization of (6.3) are a correct description of the problem of determining the β_{tj}, only if $2q+1 \leqq t \leqq n$. See also the text below (6.10). In Appendix B the validity of (6.18) is proved for the case that $2 \leqq t \leqq 2q$.

2.7. The matrix P in the case of ARMA(p,q) disturbances

2.7.1 A derivation of the formula for the autocovariance matrix of an ARMA(p,q) process

Now we assume the disturbances u_t of a time series regression model to follow a stationary ARMA(p,q) process

$$\varphi(B)u_t = \lambda(B)\varepsilon_t \qquad (7.1)$$

where ε_t is (iid) N(0,1), and $\varphi(B)$ and $\lambda(B)$ are lag polynomials

$$\varphi(B) = 1 - \varphi_1 B - \varphi_2 B^2 - \ldots - \varphi_p B^p \qquad (7.2a)$$

$$\lambda(B) = 1 - \lambda_1 B - \lambda_2 B^2 - \ldots - \lambda_q B^q \qquad (7.2b)$$

with B being the backward shift operator.

Before paying attention to an appropriate transformation matrix P we will derive explicit formulae for the autocovariances γ_k. Next it will be shown how the matrix P can be constructed if the γ_k are known. Define $\nu_k = Eu_t\varepsilon_{t-k}$ and write (7.1) alternatively as

$$u_t - \varphi_1 u_{t-1} - \dots - \varphi_p u_{t-p} = \varepsilon_t - \lambda_1 \varepsilon_{t-1} - \dots - \lambda_q \varepsilon_{t-q} \qquad (7.3)$$

Then upon multiplying throughout (7.3) by ε_{t-k} and taking expectations for $k=0,1,\dots,q$ successively, we obtain the following q+1 equations

$$\begin{bmatrix} 1 & 0 & \cdots & & 0 \\ -\varphi_1 & 1 & & & \cdot \\ \cdot & \cdot & & & \cdot \\ \cdot & & \cdot & & \cdot \\ -\varphi_{q-1} & & \cdots & 1 & 0 \\ -\varphi_q & & \cdots & -\varphi_1 & 1 \end{bmatrix} \begin{bmatrix} \nu_0 \\ \nu_1 \\ \cdot \\ \cdot \\ \cdot \\ \nu_q \end{bmatrix} = \begin{bmatrix} 1 \\ -\lambda_1 \\ \cdot \\ \cdot \\ \cdot \\ -\lambda_q \end{bmatrix} \qquad (7.4)$$

As a matter of fact we have $\varphi_{p+1} = \dots = \varphi_q = 0$, if $q>p$. The ν_k can easily be calculated from (7.4) ($k=0,1,\dots,q$). It should be noted that $\nu_k=0$ if $k<0$; see also Ljung and Box (1979).

Define $s=\max(p,q)$ and let $0_{(i,j)}$ denote an ixj matrix of zeroes, then upon multiplying (7.3) by u_{t-k} for $k=0,1,\dots,s$ and taking expectations we obtain the following Box Jenkins (1976, p. 74) equations

$$\gamma_k = \varphi_1\gamma_{k-1} + \dots + \varphi_p\gamma_{k-p} + \nu_{-k} - \lambda_1\nu_{-k+1} - \dots - \lambda_q\nu_{-k+q}, \quad k=0,1,\dots s \qquad (7.5a)$$

or in obvious matrix notation

$$\begin{bmatrix} \gamma_0 \\ \gamma_1 \\ \cdot \\ \cdot \\ \cdot \\ \gamma_p \\ \cdot \\ \cdot \\ \cdot \\ \gamma_s \end{bmatrix} = \begin{bmatrix} \gamma_1 & \gamma_2 & \cdots & \gamma_p \\ \gamma_0 & \gamma_1 & \cdots & \gamma_{p-1} \\ \cdot & & & \cdot \\ \cdot & & & \cdot \\ \cdot & & & \cdot \\ \gamma_{p-1} & \cdots & & \gamma_0 \\ \cdot & & & \cdot \\ \cdot & & & \cdot \\ \cdot & & & \cdot \\ \gamma_{s-1} & \cdots & & \gamma_{s-p} \end{bmatrix} \begin{bmatrix} \varphi_1 \\ \varphi_2 \\ \cdot \\ \cdot \\ \varphi_p \end{bmatrix} + \begin{bmatrix} 1 & -\lambda_1 & \cdots & -\lambda_q \\ -\lambda_1 & \cdots & -\lambda_q & 0 \\ \cdot & & \cdot & \cdot \\ \cdot & \cdot & & \cdot \\ \cdot & & & \\ -\lambda_q & 0 & \cdots & 0 \\ \hdashline & & & \\ & 0_{(s-q,q+1)} & & \end{bmatrix} \begin{bmatrix} \nu_0 \\ \nu_1 \\ \cdot \\ \cdot \\ \cdot \\ \nu_q \end{bmatrix} \tag{7.5b}$$

McLeod (1975) provides an algorithm, although not entirely correct, for solving these equations. The correction can be found in McLeod (1977). See also Ansley (1980) and Kohn and Ansley (1982). In the sequel of this subsection we will derive an analytical expression for the autocovariances γ_k. As regards the first p+1 equations of (7.5b), the left hand part on the right hand side of (7.5b) can be rewritten as

$$\begin{bmatrix} \gamma_1 & \gamma_2 & \cdots & \gamma_p \\ \gamma_0 & \gamma_1 & \cdots & \gamma_{p-1} \\ \cdot & & & \cdot \\ \cdot & & & \cdot \\ \cdot & & & \cdot \\ \gamma_{p-2} & & \cdots & \gamma_1 \\ \gamma_{p-1} & & \cdots & \gamma_0 \end{bmatrix} \begin{bmatrix} \varphi_1 \\ \cdot \\ \cdot \\ \cdot \\ \varphi_p \end{bmatrix} = \begin{bmatrix} 0 \\ \varphi_1\gamma_0 \\ \varphi_1\gamma_1+\varphi_2\gamma_0 \\ \cdot \\ \cdot \\ \cdot \\ \varphi_1\gamma_{p-1}+\cdots+\varphi_p\gamma_0 \end{bmatrix} + \begin{bmatrix} \varphi_1\gamma_1+\cdots+\varphi_p\gamma_p \\ \cdot \\ \cdot \\ \cdot \\ \varphi_{p-1}\gamma_1+\varphi_p\gamma_2 \\ \varphi_p\gamma_1 \\ 0 \end{bmatrix}$$

$$= \begin{bmatrix} 0 & 0 & \cdots & 0 & 0 \\ \varphi_1 & 0 & & \cdot & \cdot \\ \varphi_2 & \varphi_1 & & \cdot & \cdot \\ \cdot & \cdot & & \cdot & \cdot \\ \cdot & & \cdot & & \\ \cdot & & \cdot & 0 & 0 \\ \varphi_p & \cdots & \varphi_2 & \varphi_1 & 0 \end{bmatrix} \begin{bmatrix} \gamma_0 \\ \gamma_1 \\ \cdot \\ \cdot \\ \cdot \\ \gamma_p \end{bmatrix} + \begin{bmatrix} 0 & \varphi_1 & \varphi_2 & \cdots & \varphi_p \\ 0 & \varphi_2 & \cdots & \varphi_p & 0 \\ \cdot & \cdot & & \cdot & \cdot \\ \cdot & \cdot & \cdot & & \cdot \\ 0 & \varphi_p & & & \cdot \\ 0 & 0 & \cdots & & 0 \end{bmatrix} \begin{bmatrix} \gamma_0 \\ \gamma_1 \\ \cdot \\ \cdot \\ \cdot \\ \gamma_p \end{bmatrix} \tag{7.6}$$

Define the (p+1)×1 vector γ as $\gamma=(\gamma_0,\gamma_1,\ldots,\gamma_p)'$, then (7.6) can be denoted as $(A+B)\gamma$. Concerning the last m-p equations of (7.5b) a similar operation of interchanging the γ_k and φ_j can be performed. This yields

$$\begin{bmatrix} \gamma_p & \gamma_{p-1} & \cdots & \gamma_1 \\ \gamma_{p+1} & \gamma_p & \cdots & \gamma_2 \\ \cdot & & & \cdot \\ \cdot & & & \cdot \\ \cdot & & & \cdot \\ \gamma_{s-1} & \cdots & & \gamma_{s-p} \end{bmatrix} \begin{bmatrix} \varphi_1 \\ \cdot \\ \cdot \\ \cdot \\ \cdot \\ \varphi_p \end{bmatrix} = \begin{bmatrix} 0 & \varphi_p & \varphi_{p-1} & \cdots & \varphi_1 & 0 & \cdots & 0 \\ 0 & 0 & \varphi_p & & & \varphi_1 & & \cdot \\ \cdot & & & \cdot & & \cdot & & \cdot \\ \cdot & & & & \cdot & & \cdot & \cdot \\ \cdot & & & & & \cdot & \cdot & 0 \\ 0 & \cdots & & 0 & \varphi_p & \cdots & & \varphi_1 \end{bmatrix} \begin{bmatrix} \gamma_0 \\ \gamma_1 \\ \cdot \\ \cdot \\ \cdot \\ \gamma_{s-1} \end{bmatrix} \tag{7.7}$$

The ν_k in (7.5b) can be replaced by their solution from (7.4). Now using (7.6) and (7.7) the γ_k can be determined from (7.5b) which yields

$$\begin{bmatrix} \gamma_0 \\ \gamma_1 \\ \cdot \\ \cdot \\ \cdot \\ \gamma_p \\ \gamma_{p+1} \\ \cdot \\ \cdot \\ \cdot \\ \gamma_s \end{bmatrix} = \left[\begin{array}{ccccccccc} & & A + B - I_{p+1} & & \vdots & & 0_{(p+1,s-p)} & & \\ \hline 0 & \varphi_p & \cdots & \varphi_1 & -1 & 0 & \cdots & & 0 \\ \cdot & & & & & \cdot & & & \cdot \\ \cdot & & & & & & \cdot & & \cdot \\ \cdot & & & & & & & -1 & 0 \\ 0 & \cdots & 0 & \varphi_p & \cdots & & & \varphi_1 & -1 \end{array}\right]^{-1} \times$$

$$\times \left[\begin{array}{ccccc} -1 & \lambda_1 & \cdots & & \lambda_q \\ \lambda_1 & \cdots & & \lambda_q & 0 \\ \cdot & & \cdot & & \cdot \\ \cdot & \cdot & & & \cdot \\ \cdot & & & & \cdot \\ \lambda_q & 0 & \cdots & & 0 \\ \hline & & 0_{(s-q,q+1)} & & \end{array}\right] \begin{bmatrix} -1 & 0 & \cdots & & 0 \\ \varphi_1 & -1 & & & \cdot \\ \cdot & \cdot & \cdot & & \\ \cdot & \cdot & \cdot & -1 & 0 \\ \varphi_q & \cdots & & \varphi_1 & -1 \end{bmatrix}^{-1} \begin{bmatrix} -1 \\ \lambda_1 \\ \cdot \\ \cdot \\ \lambda_q \end{bmatrix} \tag{7.8}$$

If $k>s$, then the γ_k follow from multiplying (7.3) by u_{t-k} and taking expectations. Because $\nu_k=0$ for $k<0$ we obtain

$$\gamma_k = \varphi_1\gamma_{k-1} + \dots + \varphi_p\gamma_{k-p}, \quad k>s \tag{7.9}$$

Hence, the γ_k can be determined recursively from (7.8) and (7.9) for k>s. Finally, for k>q the ν_k follow from (7.3) after multiplying by ε_{h-k} and taking expectations, which yields

$$\nu_k = \varphi_1\nu_{k-1} + \dots + \varphi_p\nu_{k-p}, \quad k>q \tag{7.10}$$

This means that the ν_k can be determined recursively from (7.4) and (7.10) (k=0,1,2,...). An expression similar to (7.8) is given by Ljung and Box (1979) for $\gamma_0,\gamma_1,\dots,\gamma_p$. However they do not explain the form A+B in (7.8), that is, they do not mention (7.6) which equation is certainly not trivial.

In order to arrive at the same result as given by Ljung and Box denote (7.8) by

$$\begin{bmatrix} a_1 \\ a_2 \end{bmatrix} = \begin{bmatrix} M_{11} & 0 \\ M_{21} & M_{22} \end{bmatrix}^{-1} \begin{bmatrix} b_1 \\ b_2 \end{bmatrix} \tag{7.11}$$

Then using the well-known expression for the inverse of a partitioned matrix we get

$$a_1 = (M_{11})^{-1}b_1 \tag{7.12a}$$

and

$$a_2(r) = \gamma_{r-1} = \varphi_1\gamma_{r-2} + \dots + \varphi_p\gamma_{r-p-1} - b_2(r), \quad r=p+2,\dots,s+1 \tag{7.12b}$$

where use has been made of the specific structure of M_{21} and M_{22}.

Furthermore, it emerges from (7.9) and (7.10) that the γ_k and ν_k obey the same homogeneous difference equation. Only the boundary conditions for calculating the corresponding constants are different as can be seen from

(7.4) and (7.8). Because of this similarity we only mention the general solution of (7.9). Let $z_1, z_2, \ldots, z_p$ be the roots of $\varphi(1/B)=0$, then the γ_k can be expressed as

$$\gamma_k = \sum_{r=1}^{p} c_r \xi_{rk}, \quad k \geq s+1 \tag{7.13}$$

The ξ_{rk} follow from the roots $z_1, z_2, \ldots, z_p$ in the same manner as $\psi_{1j}, \ldots, \psi_{qj}$ in (6.13) do from the roots of (6.14). The p constants c_r can be determined from the values of $\gamma_{s-p+1}, \gamma_{s-p+2}, \ldots, \gamma_s$ which follow from (7.8). Solving these p boundary conditions yields

$$\begin{bmatrix} c_1 \\ \cdot \\ \cdot \\ \cdot \\ c_p \end{bmatrix} = \begin{bmatrix} \xi_{1,s-p+1} & \cdots & \xi_{p,s-p+1} \\ \cdots & & \\ \xi_{1s} & \cdots & \xi_{ps} \end{bmatrix}^{-1} \begin{bmatrix} \gamma_{s-p+1} \\ \cdot \\ \cdot \\ \cdot \\ \gamma_s \end{bmatrix} \tag{7.14}$$

Substitution of (7.14) into (7.13) gives

$$\gamma_k - (\zeta_{1k}, \ldots, \zeta_{pk}) \begin{bmatrix} \xi_{1,s-p+1} & \cdots & \xi_{p,s-p+1} \\ \cdots & & \\ \xi_{1s} & \cdots & \xi_{ps} \end{bmatrix}^{-1} \begin{bmatrix} \gamma_{s-p+1} \\ \cdot \\ \cdot \\ \cdot \\ \gamma_s \end{bmatrix}, \quad k>s-p \tag{7.15}$$

Hence, (7.15) in combination with (7.8) determines all γ_k, $k=0,1,\ldots,n$.

As regards the ν_k it is worthwhile to notice that these coefficients are equal to the π_k, where the π_k are defined as the moving average parameters in

$$u_t = \sum_{k=0}^{\infty} \pi_k \varepsilon_{t-k} \tag{7.16}$$

This follows immediately from multiplying both sides of (7.16) by ε_{t-j} and taking expectations which yields $\nu_j = \pi_j$; see also McLeod (1975).

Different formulae for the autocovariance matrix of an ARMA process are given in Zinde-Walsh (1988); her derivation is based on spectral analysis.

2.7.2 The matrix P in the case of ARMA(p,q) disturbances

Now we pay attention to the derivation of an appropriate transformation matrix P if the u_t follow a stationary ARMA(p,q) process. In the derivation two steps can be distinguished. The first step starts with a transformation of the u_t according to

$$z_t = u_t, \quad 1 \leqq t \leqq p \tag{7.17a}$$

$$z_t = -\sum_{k=0}^{p} \varphi_k u_{t-k}, \quad p+1 \leqq t \leqq n \tag{7.17b}$$

where $\varphi_0=-1$. Then the z_t follow an MA(q) process for $t \geqq p+1$. Therefore, an $(n-p)\times(n-p)$ transformation matrix R can be determined from (3.9), (4.12) and (6.18). Alternatively, the matrix R can be calculated recursively by employing the Levinson-Durbin formulae given in Section 4. Since $\gamma_k=0$ for $k>q$, this calculation will not be time consuming. However, when applying the LD formulae directly for the calculation of the matrix P in the case of an ARMA(p,q) process the numerical stability may be poor because of rounding errors, particularly when the process approaches nonstationarity; see Section 4.2 of this chapter. Let r_{tm} be an element of R for $p+1 \leqq t,m \leqq n$. The matrix R is defined in such a way that

$$Ev_t^2 = 1, \quad t \geqq p+1 \tag{7.18a}$$

$$Ev_t v_m = 0, \quad t \neq m \tag{7.18b}$$

where

$$v_t = \sum_{m=p+1}^{t} r_{tm} z_m, \quad t \geqq p+1 \tag{7.18c}$$

Upon substituting (7.17b) into (7.18c) we can write v_t as

$$v_t = - \sum_{m=p+1}^{t} \sum_{k=0}^{p} r_{tm}\varphi_k u_{m-k}$$

$$= - \sum_{m=p+1}^{t} \sum_{j=m-p}^{m} r_{tm}\varphi_{m-j}u_j, \quad t\geqq p+1 \tag{7.19}$$

Now the elements of the last n-p rows of P can be determined from (7.19)

$$p_{tj} = - \sum_{m=\max(p+1,j)}^{\min(p+j,t)} r_{tm}\varphi_{m-j}, \quad t\geqq p+1 \text{ and } j\leqq t \tag{7.20}$$

It should be noted that $p_{tj}=0$ if $j>t$.

In the second step the elements of the first p rows of P are determined by a recursive method which is similar to the recursive method described in Section 4. We assume that the elements p_{mj} have already been calculated ($m=1,2,\ldots,t-1$; $j=1,2,\ldots,n$). Thus, the orthonormal v_m can be written as

$$v_m = \sum_{j=1}^{n} p_{mj}u_j, \quad m=1,2,\ldots,t-1,p+1,\ldots,n \tag{7.21}$$

Then we define F_0 as the subspace which is spanned by $v_{p+1},\ldots,v_n$ and for $t=1,2,\ldots,p$ we define F_t as the subspace which is spanned by $v_1,\ldots,v_t,v_{p+1}$, $\ldots,v_n$. Furthermore we define w_t as the difference between u_t and its projection onto the subspace F_{t-1} and we write

$$w_t = u_t - \sum_{m\in I_t} d_{tm}v_m \tag{7.22}$$

where the index set I_t is defined as

$$I_1 = \{p+1,\ldots,n\}$$

$$I_t = \{1,2,\ldots,t-1,p+1,\ldots,n\}, \qquad t=2,3,\ldots,p.$$

Since the v_m are orthonormal for $m\in I_t$ and because of (7.21) we have

$$d_{tm} = Eu_t v_m$$
$$= \sum_{j=1}^{n} p_{mj}\gamma_{|j-t|}, \quad m\in I_t \tag{7.23}$$

It should be noted that p_{mj} in (7.23) is equal to zero, if j>m>p. Furthermore, from (4.7) it can be seen that σ_t, the length of w_t, is equal to

$$\sigma_t = (\gamma_0 - \sum_{m\in I_t} d_{tm}^2)^{1/2} \tag{7.24}$$

If we now define $v_t=w_t/\sigma_t$, then upon substituting (7.21) into (7.22) we can write v_t as

$$v_t = \frac{1}{\sigma_t}u_t - \frac{1}{\sigma_t}\sum_{m\in I_t}\sum_{j=1}^{n} d_{tm}p_{mj}u_j, \quad t=1,\dots,p \tag{7.25}$$

From (7.25) it follows that the elements of the tth row of P are equal to

$$p_{tj} = \frac{\delta_{tj}}{\sigma_t} - \frac{1}{\sigma_t}\sum_{m\in I_t} d_{tm}p_{mj}, \quad t=1,\dots,p \tag{7.26}$$

where d_{tm} and σ_t are determined from (7.23) and (7.24), respectively. The δ_{tj} in (7.26) stands for the Kronecker delta. From (7.26) it can be seen that row t of P can be determined for successively t=1,2,...,p. It should be noted that the procedure which is described in this section, will produce the usual lower triangular transformation matrix if the u_t follow an AR(p) process.

Next we will derive an expression for the determinant of the transformation matrix P in the case of an ARMA(p,q) process. According to (4.17) we have $|P|=\{V(u_1,\dots,u_n)\}^{-1}$. Because the Jacobian of transformation (7.17) is unity, we can also write

$$|P| = \{V(z_1,\ldots,z_n)\}^{-1}. \tag{7.27}$$

Since σ_t, defined in (7.24), can be seen as the length of the perpendicular between z_t and its projection onto F_{t-1} (t=1,2,...,p), and since $(r_{tt})^{-1}$ can be seen as the length of the perpendicular between z_t and its projection onto the subspace spanned by $z_{p+1},\ldots,z_{t-1}$ (t=p+2,...,n), we have

$$V(z_1,\ldots,z_n) = \prod_{t=1}^{p} \sigma_t \prod_{t=p+1}^{n} (r_{tt})^{-1}. \tag{7.28}$$

Further, it follows from (7.20) that $p_{tt}=r_{tt}$. Hence, combination of (7.27) and (7.28) gives the Jacobian

$$|P| = \prod_{t=1}^{p} \sigma_t^{-1} \prod_{t=p+1}^{n} p_{tt} \tag{7.29}$$

Finally, we will pay some attention to seasonal processes. In empirical work one is often faced by a seasonal ARMA$(p,q)_s$ process among the disturbances, in particular if the estimation of the regression model is based on monthly (s=12) or quarterly (s=4) data. The ARMA$(p,q)_s$ process can be written as

$$\varphi(B^s)u_t = \lambda(B^s)\varepsilon_t, \quad 1\leqq t\leqq n \tag{7.30}$$

where $\varphi(B^s)$ and $\lambda(B^s)$ are polynomials in B^s of degree p and q, respectively. Further, we introduce the auxiliary ARMA(p,q) process

$$\varphi(B)z_t = \lambda(B)\varepsilon_t, \quad 1\leqq t\leqq m \tag{7.31}$$

Define $\gamma_k=Eu_tu_{t-k}$ and $\sigma_k=Ez_tz_{t-k}$, then the following relationship between the autocovariances of (7.30) and (7.31) can easily be established

$$\gamma_{ks} = \sigma_k \tag{7.32}$$

For simplicity and without loss of generality we assume n=ms. Then using the Kronecker product notation we can write (7.32) as $\Gamma_n=\Sigma_m \otimes I_s$, Γ_n and Σ_m being the autocovariance matrices of (7.30) and (7.31), respectively. Hence, an appropriate transformation matrix for the seasonal ARMA$(p,q)_s$ process is $P=Q\otimes I_s$, where Q is such that $Q\Sigma_m Q'=I_m$. For the Jacobian we have $|P|=|Q|^s$; see Section 2. If we suppose $(m-1)s<n<ms$, then define $r=ms-n$. To obtain now an appropriate transformation matrix the last r columns and rows of $Q\otimes I_s$ have to be deleted. If we assume Q to be lower triangular or to be obtained by applying the method described in this section, then the Jacobian in this case is equal to $|Q|^s/q_{mm}^r$.

Appendix 2.A Linear vector spaces

It is recalled from linear algebra (see, e.g., Friedman (1982) and Shilov (1977)) that a non-empty set S of elements is called a linear vector space over the real number field **R** if

A) There is an addition rule with the following properties

i) $x,y\in S \rightarrow x+y\in S$;

ii) $x+y=y+x$ for every $x,y\in S$;

iii) $(x+y)+z=x+(y+z)$ for every $x,y,z\in S$;

iv) there exists an element $0\in S$ (the zero vector) such that $x+0=x$ for every $x\in S$;

v) for every $x\in S$ there exists an element $y\in S$ such that $x+y=0$.

B) There is a multiplication rule with the following properties

vi) $x\in S \rightarrow \lambda x\in S$ for every $\lambda\in R$;

vii) $1.x=x$ for every $x\in S$;

viii) $\alpha(\beta x)=(\alpha\beta)x$ for every $x\in S$ and every $\alpha,\beta\in R$;

ix) $(\alpha+\beta)x=\alpha x+\beta x$ for every $x\in S$ and every $\alpha,\beta\in R$;

x) $\alpha(x+y)=\alpha x+\alpha y$ for every $x,y\in S$ and every $\alpha\in R$.

A linear vector space is called Euclidean or an inner product space if

C) There is a rule assigning to every pair of elements (vectors) $x,y\in S$ a real number called the inner product (scalar product) of x and y, denoted by (x,y), such that

xi) $(x,y)=(y,x)$ for every $x,y\in S$;

xii) $(x,y+z)=(x,y)+(x,z)$ for every $x,y,z\in S$;

xiii) $(\lambda x,y)=\lambda(x,y)$ for every $x,y\in S$ and every $\lambda\in R$;

xiv) $(x,x)>0$ for every $x\neq 0$ and $(x,x)=0$ for $x=0$.

The length or norm of a vector x in the Euclidean space S is defined by

xv) $\|x\|=(x,x)^{1/2}$

while the angle between two vectors x and y is defined by the angle (lying between 0 and 180 degrees) whose cosine is equal to

xvi) $\cos(x,y) = \dfrac{(x,y)}{\|x\|\cdot\|y\|}$.

The elements of S can be called vectors, regardless of the fact that their concrete nature may be quite unlike the more familiar directed line segments. Since some authors prefer the term Hilbert space, the definition of a Hilbert space is also given. A set S with the properties i)-xvi) is called a Hilbert space if

xvii) for every Cauchy sequence $\{x_n\}\subset S$ with $\lim_{m,n\to\infty}(x_n-x_m,x_n-x_m)=0$ there exists an $x\in S$ such that $\lim_{n\to\infty}(x_n-x,x_n-x)=0$.

For the complex case, i.e. R is replaced by the complex number field C and (x,y) is a complex valued function, there is the additional condition

xviii) $(x,y)=\overline{(y,x)}$ for every $x,y,\in S$.

Furthermore, if the definition of the norm $\|x\|$ is different from xv) then S is called a Banach space. However, the norm which is a real valued function and denoted by $\|x\|$, should always have the following properties

ixx) $\|x\|=0$ if and only if $x=0$;

xx) $\|\lambda x\|=|\lambda|.\|x\|$ for every $x\in S$ and $\lambda\in R$ (or $\lambda\in C$);

xxi) $\|x+y\|\leq\|x\|+\|y\|$ for every $x,y\in S$.

Appendix 2.B The formula for β_{tj} if t is small

In this appendix we prove the validity of (6.18) for $t=2,3,\ldots,2q$. For that purpose we explicitly write the 2q boundary conditions which have to be solved in order to obtain the 2q constants if $2q<t\leq n$. The first q boundary conditions are according to (6.3)

$$\begin{bmatrix} \gamma_0 & \cdots & \gamma_{q-1} & \gamma_q & 0 & \cdots & 0 \\ \cdot & & \cdot & \cdot & & & \cdot \\ \cdot & & \cdot & \cdot & & & \cdot \\ \cdot & & \cdot & \cdot & & & 0 \\ \gamma_{q-1} & \cdots & \gamma_0 & \gamma_1 & \cdots & & \gamma_q \end{bmatrix} \begin{bmatrix} \beta_{t1} \\ \cdot \\ \cdot \\ \cdot \\ \beta_{t,2q} \end{bmatrix} = \begin{bmatrix} \gamma_1 \\ \cdot \\ \cdot \\ \cdot \\ \gamma_q \end{bmatrix} \tag{B.1}$$

The last q boundary conditions are according to (6.3)

$$\begin{bmatrix} \gamma_q & \cdots & \gamma_1 & \gamma_0 & \cdots & \gamma_{q-1} \\ 0 & & \cdot & \cdot & & \cdot \\ \cdot & & \cdot & \cdot & & \cdot \\ \cdot & & \cdot & \cdot & & \cdot \\ 0 & \cdots & 0 & \gamma_q & \gamma_{q-1} \cdots & \gamma_0 \end{bmatrix} \begin{bmatrix} \beta_{t,t-2q} \\ \cdot \\ \cdot \\ \cdot \\ \beta_{t,t-1} \end{bmatrix} = \begin{bmatrix} 0 \\ \cdot \\ \cdot \\ \cdot \\ 0 \end{bmatrix} \tag{B.2}$$

The generalization of (6.9) implies

$$\beta_{tt} = \ldots = \beta_{t,t+q-1} = 0 \tag{B.3}$$

Now we first prove the validity of (6.18) in the case where $2\leq t\leq q$.

Substituting (B.3) into (B.1) we obtain the normal equations corresponding to $\beta_{t1},\ldots,\beta_{t,t-1}$

$$\begin{bmatrix} \gamma_0 & \gamma_1 & \cdots & \gamma_{t-2} \\ \gamma_1 & \gamma_0 & & \cdot \\ \cdot & & & \cdot \\ \cdot & & & \cdot \\ \gamma_{t-2} & \cdots & & \gamma_0 \end{bmatrix} \begin{bmatrix} \beta_{t1} \\ \cdot \\ \cdot \\ \cdot \\ \beta_{t,t-1} \end{bmatrix} = \begin{bmatrix} \gamma_1 \\ \cdot \\ \cdot \\ \cdot \\ \gamma_{t-1} \end{bmatrix} \tag{B.4}$$

Secondly, we consider the case where $q<t\leqq 2q$. According to (4.11) the β_{tj} have to satisfy the following t-1 normal equations

$$\begin{bmatrix} \gamma_0 & \gamma_1 & \cdots & \gamma_q & 0 & \cdots & 0 \\ \gamma_1 & \gamma_0 & & & \cdot & & \cdot \\ \cdot & & & & & \cdot & \cdot \\ \cdot & & & & & \cdot & 0 \\ \cdot & & & & & & \gamma_q \\ \gamma_q & & & & & & \cdot \\ 0 & \cdot & & & & & \cdot \\ \cdot & & \cdot & & & & \cdot \\ \cdot & & & \cdot & & & \cdot \\ 0 & \cdots & 0 & \gamma_q & \cdots & & \gamma_0 \end{bmatrix} \begin{bmatrix} \beta_{t1} \\ \\ \\ \cdot \\ \cdot \\ \cdot \\ \\ \\ \\ \beta_{t,t-1} \end{bmatrix} = \begin{bmatrix} \gamma_1 \\ \cdot \\ \cdot \\ \cdot \\ \gamma_q \\ 0 \\ \cdot \\ \cdot \\ \cdot \\ 0 \end{bmatrix} \tag{B.5}$$

The first q equations of (B.5) immediately follow from the substitution of (B.3) into the q boundary conditions of (B.1), while the last t-q-1 equations of (B.5) are exactly equal to the last t-q-1 boundary conditions of (B.2), which establishes what is required and concludes the proof.

CHAPTER III

COMPUTATIONAL ASPECTS OF DATA TRANSFORMATIONS AND ANSLEY'S ALGORITHM

3.1 Introduction

In Section 5 of the previous chapter a simple recursive algorithm was given for an appropriate transformation of the disturbances and the data of a regression model when the disturbances follow an MA(1) process. In the Sections 2 and 3 of this chapter auxiliary recursive formulae are derived in order to transform the disturbances u_t as well as the data when the disturbances follow an arbitrary MA or ARMA process, and where the number of multiplications involved only increases linearly with the number of observations. Furthermore, it will be shown in this chapter in Section 4 how the method of Ansley (1979) in the case of ARMA(p,q) disturbances can be put in the framework of the previous chapter and how this leads to an algorithm for calculating $|\Gamma_u|$ as well as $v=L^{-1}z$ where L is the lower triangular matrix such that $LL'=\Gamma_z$, i.e., the Cholesky decomposition; z is defined in Section 7 of the previous chapter.

In the sequel of this chapter the equations of Chapter 2 which are referred to, are indicated with three distinct numbers.

3.2 Recursive computations for models with MA(q) disturbances

Consider the MA(q) process. It can be seen from (2.3.1c), (2.3.9) and (2.6.13) that the appropriate transformation can be written as

$$v_t = \sum_{j=1}^{t} p_{tj} u_j = \frac{1}{\sigma_t}(u_t - \sum_{j=1}^{t-1} \beta_{tj} u_{t-j})$$

$$= \frac{1}{\sigma_t}(u_t - \sum_{j=1}^{t-1} \sum_{p=1}^{2q} c_{tp} \psi_{pj} u_{t-j}) \qquad (2.1)$$

Define $U_{pt} = \sum_{j=1}^{t-1} \psi_{pj} u_{t-j}$, $p=1,2,\ldots,2q$. Then (2.1) may be rewritten

$$v_t = \frac{1}{\sigma_t}(u_t - \sum_{p=1}^{2q} c_{tp} U_{pt}) \qquad (2.2)$$

Now the U_{pt} can be recursively calculated as follows. Three relevant cases have to be distinguished; see Section 6 of the previous chapter for the symbols used, and their meaning. Firstly, assume x_p is a unique real valued root of (2.6.14). Then $\psi_{pj} = x_p^j$ and we have

$$U_{p,t+1} = \sum_{j=1}^{t} x_p^j u_{t+1-j}$$

$$= x_p u_t + \sum_{j=2}^{t} x_p^j u_{t-j+1}$$

$$= x_p u_t + \sum_{j=1}^{t-1} x_p^{j+1} u_{t-j}$$

$$= x_p u_t + x_p U_{pt} \qquad (2.3)$$

Secondly, assume x_1 is a repeated real valued root of (2.6.14) of multiplicity m. Then we have $\psi_{pj} = j^{p-1} x_1^j$ $(p=1,2,\ldots,m)$. Now we obtain

$$U_{p,t+1} = \sum_{j=1}^{t} j^{p-1} x_1^j u_{t+1-j}$$

$$= x_1 u_t + \sum_{j=2}^{t} j^{p-1} x_1^j u_{t+1-j}$$

$$= x_1 u_t + \sum_{j=1}^{t-1} (j+1)^{p-1} x_1^{j+1} u_{t-j}$$

$$= x_1 u_t + x_1 \sum_{j=1}^{t-1} \sum_{s=1}^{p} \binom{p-1}{s-1} j^{s-1} x_1^j u_{t-j}$$

$$= x_1 u_t + x_1 \sum_{s=1}^{p} \binom{p-1}{s-1} U_{st} \tag{2.4}$$

Thirdly, assume x_1 is a repeated complex valued root of (2.6.14) of multiplicity m. Set $x_1 = x\exp(i\varphi)$. Then we have $\psi_{pj} = j^{p-1}x^j\sin(j\varphi)$ and $\psi_{p+m,j} = j^{p-1}x^j\cos(j\varphi)$ $(p=1,2,\ldots,m)$. Employing the well-known trigonometric formulae we obtain

$$U_{p,t+1} = x\sin(\varphi)\,u_t + \sum_{j=2}^{t} j^{p-1}\,x^j\sin(j\varphi)\,u_{t+1-j}$$

$$= x\sin(\varphi)\,u_t + \sum_{j=1}^{t-1} (j+1)^{p-1}x^{j+1}\sin\{(j+1)\varphi\}\,u_{t-j}$$

$$= xu_t\sin(\varphi) + x \sum_{j=1}^{t-1} \sum_{s=1}^{p} \binom{p-1}{s-1} j^{s-1}x^j\{\sin(j\varphi)\cos(\varphi)+\cos(j\varphi)\sin(\varphi)\}u_{t-j}$$

$$= xu_t\sin(\varphi) + x \sum_{s=1}^{p} \binom{p-1}{s-1} \{U_{st}\cos(\varphi) + U_{s+m,t}\sin(\varphi)\} \tag{2.5}$$

Similarly, we have

$$U_{p+m,t+1} = xu_t\cos(\varphi) + x \sum_{s=1}^{p} \binom{p-1}{s-1} \{U_{s+m,t}\cos(\varphi) - U_{st}\sin(\varphi)\} \tag{2.6}$$

Further, it follows from (2.4.12) and (2.6.13)

$$\sigma_t^2 = \gamma_0 - \sum_{j=1}^{q} \sum_{p=1}^{3q} c_{tp}\psi_{pj}\gamma_j, \quad t=1,2,\ldots,n \tag{2.7}$$

Therefore, it may be concluded that the number of multiplications for calculating the appropriately transformed data is proportional to n, with n being the number of observations.

3.3 Recursive computations for models with ARMA(p,q) disturbances

Next, consider the ARMA(p,q) process. Transformation (2.7.17) can easily be performed and transformation (2.7.18c) is like the transformation of MA(q) disturbances given above. The only problem is transformation (2.7.22) which requires the calculation of the d_{tm}, $m \in I_t$ $(t=1,2,\ldots,p)$. Given the d_{tm}, the σ_t can be calculated from (2.7.24) and we obtain $v_t = w_t/\sigma_t$ $(t=1,2,\ldots,p)$. If $m>p$ then by employing (2.7.18c) the d_{tm} can be calculated in a computationally attractive manner as follows

$$d_{tm} = Eu_t v_m = \sum_{s=p+1}^{m} r_{ms} Eu_t z_s, \quad m \geqq p+1 \text{ and } 1 \leqq t \leqq p \tag{3.1}$$

Define $\mu_{ts} = Eu_t z_s (s=p+1,\ldots,n)$, then from (2.7.3) and (2.7.17b) it can be seen

$$\mu_{ts} = Eu_t(\varepsilon_s - \sum_{j=1}^{q} \lambda_j \varepsilon_{s-j})$$

$$= -\sum_{j=1}^{q} \lambda_j \nu_{t-s+j}, \quad s \geqq p+1 \text{ and } 1 \leqq t \leqq p \tag{3.2}$$

See Section 7 in Chapter 2 for the calculation of the ν_t. Note that $\mu_{ts}=0$ for $s \geqq t+q$. Hence, we have for $m \geqq p+1$

$$d_{tm} = \sum_{s=p+1}^{t+q} r_{ms}\mu_{ts}, \quad \text{if } p-q<t \leqq p \tag{3.3a}$$

$$d_{tm} = 0, \quad \text{if } t \leqq p-q \tag{3.3b}$$

From (3.3a) it emerges that the first q columns of the $(n-p)\times(n-p)$ matrix R have to be calculated, which is the transformation matrix of the corresponding MA part of the ARMA process. From (2.3.9), (2.6.18) and (2.7) it can be seen that the number of operations for calculating the elements r_{ms} is increasing linearly with the number of observations, $p+1 \leqq m \leqq n$ and $p+1 \leqq s \leqq p+q$.

Finally the d_{tm} have to be calculated for $m \leq t-1$ $(t=2,3,\ldots,p)$. Since $v_m = w_m/\sigma_m$ we obtain in view of (2.7.22)

$$d_{tm} = Eu_t v_m = \frac{1}{\sigma_m} Eu_t(u_m - \sum_{s\in I_m} d_{ms} v_s)$$

$$= \frac{1}{\sigma_m}(\gamma_{t-m} - \sum_{s\in I_m} d_{ms} d_{ts}), \quad 1 \leq m \leq t-1 \tag{3.4}$$

From (3.4) it appears that d_{tm} can be recursively calculated $(m=1,2,\ldots,t-1;$ $t=2,3,\ldots,p)$. The number of multiplications involved is proportional to n.

3.4 Ansley's method

In this section a brief summary is given of the algorithm of Ansley (1979) for calculating the exact likelihood function of an ARMA(p,q) process $\{u_t\}$ and it will be demonstrated how his approach can be adopted within the framework described in Chapter 2.

Ansley's algorithm is based on the Cholesky decomposition of a symmetric band matrix, i.e., $\Gamma = LL'$ where L is the lower triangular band matrix with bandwidths corresponding to those of Γ. Both the rows of L and the solutions v_t of the equations $Lv=z$ can be determined recursively at the same time by means of an efficient algorithm; it is recalled that z is a straightforward transformation of u. The determinant $|L|$ is simply the product of its diagonal elements. For the sake of simplicity explanatory variables are omitted. Then the concentrated log likelihood function can be written as

$$\ln L(u) = \text{const.} - \frac{1}{2}n \ln SSR \tag{4.1}$$

$$SSR = \sum_{t=1}^{n} (|L|^{1/n} v_t)^2 \tag{4.2}$$

See also Section 2 of Chapter 4, viz. (4.2.15).

Now it is shown how the Cholesky decomposition of a band matrix can be carried out by means of the method presented in Chapter 2. Firstly, transform u_t following

$$z_t = u_t, \quad t=1,\dots,p \tag{4.3}$$

$$z_t = u_t - \sum_{k=1}^{p} \varphi_k u_{t-k}, \quad t>p \tag{4.4}$$

For simplicity and without loss of generality it is assumed that $p \geq q$. Secondly, it is recalled from Section 4 in Chapter 2 that L is equal to

$$L = \begin{bmatrix} \sigma_1 & 0 & \cdots & 0 \\ d_{21} & \sigma_2 & \cdots & 0 \\ \vdots & \vdots & & \vdots \\ d_{n1} & d_{n2} & \cdots & \sigma_n \end{bmatrix} \tag{4.5}$$

Apart from the first p rows and columns the autocovariance matrix Γ_z of the vector z has bandwidths equal to q and its elements are determined by

$$Ez_iz_j = Eu_iu_j = \gamma_{j-i}^{ARMA}, \quad \text{if } 1\leq i\leq j\leq p \tag{4.6}$$

$$\begin{aligned} Ez_iz_j &= Eu_iz_j \\ &= Eu_i(\varepsilon_j - \sum_{k=1}^{q} \lambda_k \varepsilon_{j-k}) \\ &= -\sum_{k=1}^{q} \lambda_k \nu_{i-j+k}, \quad \text{if } 1\leq i\leq p<j\leq p+q \end{aligned} \tag{4.7}$$

$$Ez_iz_j = Eu_iz_j = 0, \quad \text{if } 1\leq i\leq p \text{ and } j>p+q \tag{4.8}$$

$$Ez_iz_j = \gamma_{j-i}^{MA}, \quad \text{if } p<i\leq j\leq n \tag{4.9}$$

The superscripts ARMA and MA refer to the autocovariances of an ARMA(p,q) process and MA(q) process, respectively. The ν_k are given in Section 7 of Chapter 2. In the sequel of this section the elements of Γ_z are denoted by γ_{tm}. The first p elements in the vector z have an ARMA autocovariance matrix Γ_p. Therefore, the Cholesky decomposition can easily be carried out numerically such that $\Gamma_p = L_p L_p'$ and the elements in v_p follow from the solution of

$$L_p v_p = z_p \tag{4.10}$$

Then in the notation of Section 4 of Chapter 2 we have for $t \geqq p+1$

$$w_t = z_t - \sum_{k=1}^{t-1} d_{tk} v_k \tag{4.11}$$

$$d_{tk} = E z_t v_k \tag{4.12}$$

Since z_t is orthogonal to the subspace spanned by $v_1, v_2, \ldots v_{t-q-1}$, or $z_1, z_2, \ldots, z_{t-q-1}$, we have

$$d_{tk} = 0, \quad k < t-q \tag{4.13}$$

And for $t-q \leqq k < t$ we obtain in view of (2.4.4) and the first equality of (2.4.9)

$$d_{tk} = E z_t v_k = E z_t \Big(z_k - \sum_{s=k-q}^{k-1} d_{ks} v_s\Big)/\sigma_k$$

$$= \Big(\gamma_{tk} - \sum_{s=k-q}^{k-1} d_{ks} d_{ts}\Big)/\sigma_k \tag{4.14}$$

where use has been made of $d_{ks}=0$ according to (4.13), $s<k-q$. After the off-diagonal elements $\ell_{tk}=d_{tk}$ have been obtained the diagonal elements can be determined according to (2.4.7)

$$\ell_{tt} = \sigma_t = (\gamma_{tt} - \sum_{k=t-q}^{t-1} d_{tk}^2)^{½} \tag{4.15}$$

It is obvious from (4.14) and (4.15) that d_{tk} and σ_t^2 can be determined in a recursive manner. Consequently, the whole lower triangular matrix L and its determinant $|L|$ can be determined in a relatively convenient manner by employing the band structure of the covariance matrix of vector z. Then similar to (4.10) the last n-p elements in v can be determined in a recursive manner from Lv=z. Next, SSR defined by (4.2) can be calculated. This precisely is what Ansley proposes in his article. Ansley's algorithm is strongly based on the approach of Phadke and Kedem (1978). These authors consider the specific case of a multivariate MA(q) process. They propose to decomposite the corresponding autocovariance matrix Γ as $\Gamma=LDL'$, D being a diagonal matrix with elements $d_t=\sigma_t^2$. The diagonal elements of L are equal to unity whereas the off-diagonal elements of L follow from (4.14)

$$\ell_{tk} = \frac{d_{tk}}{\sigma_k} = (\gamma_{tk} - \sum_{s=k-q}^{k-1} d_{ks}d_{ts})/\sigma_k^2 \tag{4.16}$$

Using the latter representation one avoids the calculation of the square root of σ_k^2. Furthermore, Ansley refers to other authors for computationally efficient algorithms for the Cholesky decomposition of positive definite band matrices.

CHAPTER IV

GLS ESTIMATION BY KALMAN FILTERING

4.1 Introduction

Harvey and Phillips (1979) show how the framework of a state space model can be employed to find the GLS estimators of the parameters of a linear regression model when the disturbances follow an ARMA process with known parameters. The estimation can be carried out by applying the recursive Kalman filter technique for the successive observations in the periods $k+1,\ldots,n$, where k denotes the number of parameters to be estimated. By means of this approach it is also possible to find the exact maximum likelihood estimator of all the model parameters, including the parameters of the covariance matrix of the disturbances. Mélard (1983) provides a fast algorithm for calculating the exact likelihood function of an ARMA process based on the Kalman filter technique as proposed by Harvey and Phillips.

The main aim of this chapter is to give a survey of the article of Harvey and Phillips. Furthermore, we will pay attention to some other points. We will show how the exact ML method for the MA(1) case as presented by HP can be generalized for an arbitrary but stationary ARMA(p,q) process. In passing it will be demonstrated how the geometric interpretation of a linear space of stochastic variables as described in Chapter 2 can be used for the derivation of the Kalman filter equations in a relatively simple manner. More generally, the conditional distribution of a multivariate normal vector can also be derived by employing this geometric interpretation. Furthermore, it will be argued that the geometric derivations of the Kalman equations as already given by several other authors, do not directly apply to the limiting case where the initial estimate of the state vector is random rather than fixed, viz., Kalman (1960), Kailath (1981), Anderson and Moore (1979), Åström (1970), and Hannan and Deistler (1988). Also the interpretations given by Chow (1983) and Goodwin and Sin (1984) do not apply to this limiting case. Hence, these derivations are not directly very useful for the specific case that the initial estimate is random which, in general, may occur in the classical regression model. Besides, two minor errors in the derivation given by Hannan and Deistler will be discussed. The last point we pay attention to in this chapter is the derivation of a formula for a prediction interval for a sum of consecutive forecasts based on Kalman filtering.

The recursive Kalman filter equations are strongly based on standard results from multivariate analysis, therefore a brief survey of these results is given first in Section 2. Although these results are well known the derivation given here is sometimes different from the one prevailing in the literature. Next in Section 3 it is shown how the Kalman filter equations follow directly from particular results in multivariate analysis. This derivation has to my best knowledge not been mentioned elsewhere in the literature. Furthermore, a comparison is made with other derivations in the literature.

In Section 4 it is explained how the likelihood function can be calculated given an estimate of the initial state vector and its error covariance matrix.

In the Sections 5 and 6 the approach of Harvey and Phillips (HP) is presented. Although HP argue that in comparison with the exact method their approximate method performs very well in the case of an MA(1) process, it is not clear from their Monte Carlo experiments that this also is the case for the general ARMA process. Therefore an exact Kalman ML estimator for the general ARMA case is given in Section 5 which is not mentioned in the article of HP. This might be particularly useful when one of the roots of the lag polynomial of the autoregressive part of the error process is close to the unit circle. In these circumstances rounding errors may tend to build up in the recursive Kalman filter algorithm; see also Box and Jenkins (1976, p. 84). Note that the process $\{u_t\}$ explodes when one or more roots of the autoregressive lag polynomial are on or inside the unit circle while the u_t always have a finite variance when all roots are outside the unit circle. This is so, even if the roots of the corresponding MA lag polynomial are inside the unit circle.

Finally, in Section 7 a formula is derived for the prediction interval of the sum of an arbitrary number of consecutive forecasts, based on Kalman filtering.

4.2. Some results from multivariate analysis

4.2.1 Likelihood functions

Let $w_1,w_2,\ldots,w_n$ have the normal distribution with mean zero and variance σ^2 so that the density function is given by

$$g(w_t) = (2\pi\sigma^2)^{-\frac{1}{2}} e^{-\frac{1}{2}(w_t/\sigma)^2}, \quad t=1,2,\ldots,n \tag{2.1}$$

Assuming that the w_t are stochastically independent then their joint distribution is

$$g(w) = \prod_{t=1}^{n} (2\pi\sigma^2)^{-\frac{1}{2}} e^{-\frac{1}{2}(w_t/\sigma)^2}$$

$$= (2\pi\sigma^2)^{-\frac{1}{2}n} e^{\frac{-1}{2\sigma^2} w'w} \tag{2.2}$$

where w stands for the vector $(w_1,w_2,\ldots,w_n)'$. The vector w is multivariate normal with $Ew=0$ and $Eww'=\sigma^2 I$ which is denoted by

$$w \sim N(0,\sigma^2 I) \tag{2.3}$$

If a vector $u=(u_1,\ldots,u_n)'$ can be written as a linear combination of $w_1,w_2,\ldots,w_n$

$$u_t = \sum_{j=1}^{n} \ell_{tj} w_j \quad \text{or} \quad u = Lw \tag{2.4}$$

then we have $Eu=LEw=0$ and $Euu'=L(Eww')L'=\sigma^2 LL'$. Define $\Gamma=LL'$, then we can write

$$u \sim N(0,\sigma^2\Gamma) \tag{2.5}$$

Since the Jacobian of the transformation (2.4) equals

$$J = \left| \frac{\partial u}{\partial w'} \right| = |L| \tag{2.6}$$

it follows from (2.2) that the joint density function of u can be written as

$$f(u) = (2\pi\sigma^2)^{-\frac{1}{2}n} \; |L|^{-1} \; e^{-\frac{1}{2}(u'(L')^{-1}L^{-1}u)/\sigma^2} \tag{2.7}$$

$$= (2\pi\sigma^2)^{-\frac{1}{2}n} \; |\Gamma|^{-\frac{1}{2}} \; e^{-\frac{1}{2}(u'\Gamma^{-1}u)/\sigma^2} \tag{2.8}$$

The log likelihood function of the n observations $u_1,u_2,\ldots,u_n$ equals

$$\ln f(u) = -\frac{1}{2}n \ln(2\pi\sigma^2) - \frac{1}{2}\ln|\Gamma| - \frac{1}{2}(u'\Gamma^{-1}u)/\sigma^2 \tag{2.9}$$

Maximizing (2.9) with respect to σ^2 yields

$$\frac{\partial \ln f(u)}{\partial \sigma^2} = -\frac{n}{2\sigma^2} + \frac{u'\Gamma^{-1}u}{2\sigma^4} = 0 \tag{2.10}$$

or

$$\tilde{\sigma}^2 = \frac{u'\Gamma^{-1}u}{n} \tag{2.11}$$

Substitution of (2.11) into (2.9) gives the concentrated log likelihood function

$$\ln f_c(u) = -\frac{1}{2}n \ln 2\pi - \frac{1}{2}n \ln \frac{u'\Gamma^{-1}u}{n} - \frac{1}{2}\ln|\Gamma| - \frac{1}{2}n \tag{2.12}$$

In their article HP show how (2.12) can be calculated by employing the Kalman filter algorithm for the case that the disturbances u_t in a regression model follow a stationary ARMA(p,q) process given the parameters of that process.

Some other algorithms are based on the Cholesky decomposition of the covariance matrix; see Ansley (1979). Therefore, we will also give a brief derivation of the expression to be minimized in these algorithms. Define $P=L^{-1}$ then $|\Gamma|=|L|^2=|P|^{-2}$ and

$$u'\Gamma^{-1}u = u'P'Pu = w'w \tag{2.13}$$

In view of (2.13) expression (2.12) can be written as

$$\ln f_c(u) = -\frac{1}{2}n \ln 2\pi + \frac{1}{2}n \ln n - \frac{1}{2}n - \frac{1}{2}n \ln w'w + \frac{1}{2}n \ln|P|^{2/n} \tag{2.14}$$

or

$$\begin{aligned} \ln f_c(u) &= \text{const.} - \frac{1}{2}n \ln(|P|^{-2/n}w'w) \\ &= \text{const.} - \frac{1}{2}n \ln z'z \end{aligned} \tag{2.15}$$

where $z=|P|^{-1/n}w$. Hence, maximizing the log likelihood function with respect to the parameters of the covariance matrix is equivalent to the minimization of

$$\sum_{t=1}^{n} z_t^2 = z'z \tag{2.16}$$

This expression is given in Ansley (1979) and also used in Chapter 2. The quadratic form of (2.16) is due to the fact that σ^2 is unknown and has been concentrated out. Were σ^2 known then it can be seen from (2.9) that the function to be minimized becomes

$$\frac{1}{\sigma^2} u'\Gamma^{-1}u + \ln|\Gamma| \tag{2.17}$$

4.2.2 Conditional normal distributions and minimum variance estimators

Next, we consider the multivariate normal distribution of a vector y with p elements

$$y \sim N(0,\Gamma) \tag{2.18}$$

If y is partitioned into two subvectors of p_1 and p_2 elements, respectively, with $p_1+p_2=p$ then (2.18) can be written in the corresponding partitioned form

$$\begin{bmatrix} y_1 \\ y_2 \end{bmatrix} \sim N \begin{bmatrix} 0, & \Gamma_{11} & \Gamma_{12} \\ 0, & \Gamma_{21} & \Gamma_{22} \end{bmatrix} \tag{2.19}$$

Problems that now naturally arise concern the conditional distribution of y_1 given y_2 and, more generally, what is the best linear estimator of y_1 given y_2. As in regression analysis it is convenient here to write y_1 as a sum of two parts: a part linear in y_2 and a part which consists of p_1 random variables with zero expectation stacked in a residual vector u

$$y_1 = Ky_2 + u \tag{2.20}$$

In the sequel of this chapter we will call K the Kalman matrix, because this matrix plays an important role in the derivation of the Kalman filter equations as we will show in the next section. A very simple geometric interpretation of (2.20) is as follows. Denote the ith row of K by K_i. Then by the appropriate choice of the elements in K_i the stochastic variable $u_i(=y_{1i}-K_iy_2)$ in (2.20) can be seen as the orthogonal remainder, i.e., the difference of y_{1i} and its projection onto the hyperplane $H=H(y_{21},\ldots,y_{2p(2)})$ spanned by the p_2 elements in y_2; see Section 3 and Appendix A of Chapter 2 for a brief description of linear spaces of stochastic variables. In order to determine the Kalman matrix K we require u and y_2 to be stochastically independent or, equivalently, $Euy_2'=0$. Hence, postmultiplying both sides of (2.20) by y_2' and taking expectations we obtain the following normal equations

$$Ey_1y_2' = K\,Ey_2y_2' + 0 \tag{2.21}$$

or

$$\Gamma_{12} = K\Gamma_{22} \tag{2.22}$$

Assuming nonsingularity it follows from (2.22) that

$$K = \Gamma_{12}\Gamma_{22}^{-1} \tag{2.23}$$

In order to obtain the covariance matrix of y_1 given y_2 substitute (2.23) into (2.20)

$$y_1 = \Gamma_{12}\Gamma_{22}^{-1}y_2 + u \tag{2.24}$$

Given a fixed value of y_2 the covariance matrix of y_1 equals $Euu'=\Gamma_u$. Making use of $Ey_2u'=0$ we can obtain an expression for the covariance matrix of u by taking the variances of both sides of (2.24)

$$\Gamma_{11} = \Gamma_{12}\Gamma_{22}^{-1}\Gamma_{22}\Gamma_{22}^{-1}\Gamma_{21} + \Gamma_u \tag{2.25}$$

or

$$\Gamma_u = \Gamma_{11} - \Gamma_{12}\Gamma_{22}^{-1}\Gamma_{21} \tag{2.26}$$

Hence, we have $E(y_1|y_2)=\Gamma_{12}\Gamma_{22}^{-1}y_2$ and $\text{cov}(y_1|y_2)=\Gamma_{11}-\Gamma_{12}\Gamma_{22}^{-1}\Gamma_{21}$. Let now y_1 and y_2 have expectation μ_1 and μ_2, respectively. Replacing y_1 and y_2 in (2.24) by $y_1-\mu_1$ and $y_2-\mu_2$, respectively, we obtain

$$y_1 = \mu_1 + \Gamma_{12}\Gamma_{22}^{-1}(y_2-\mu_2) + u \tag{2.27}$$

Hence,

$$E(y_1|y_2) = \mu_1 + \Gamma_{12}\Gamma_{22}^{-1}(y_2-\mu_2) \tag{2.28a}$$

$$\text{cov}(y_1|y_2) = \Gamma_{11} - \Gamma_{12}\Gamma_{22}^{-1}\Gamma_{21} \tag{2.28b}$$

Summarizing the foregoing results we have

$$(y_1|y_2) \sim N(\mu_1 + \Gamma_{12}\Gamma_{22}^{-1}(y_2-\mu_2),\ \Gamma_{11} - \Gamma_{12}\Gamma_{22}^{-1}\Gamma_{21}) \tag{2.29}$$

Chow (1983, pp. 9-12) and Spanos (1986, p. 317) derive the same result in a different way.

From the point of view of estimation theory there is another interesting derivation of the Kalman matrix K. Given y_2 the question to be answered is what is the best linear estimator of the ith element of y_1 (y_{1i}); best in the least squares sense, i.e. , the variance of the estimation error $u_i(=y_{1i}-K_iy_2)$ is minimum. This leads to the following minimization problem

$$\min_{K_i} Eu_i^2, \quad i=1,2,\ldots,p_1$$

Since the rows K_i of the Kalman matrix K can be determined each separately, it follows that the sum of the minimized error variances is equal to the minimized sum of the error variances, that is

$$\sum_{i=1}^{p_1} \min_{K_i} Eu_i^2 = \min_{K} tr(Euu') = \min_{K} Eu'u = \min_{K} E\{\|u\|^2\} \tag{2.30}$$

Hence, K can be determined by minimizing tr(Euu') with respect to the elements of the Kalman matrix K. This yields

$$\begin{aligned} 0 &= \frac{\partial tr(Euu')}{\partial K} \\ &= \frac{\partial Eu'u}{\partial K} = \frac{\partial E(y_1-Ky_2)'(y_1-Ky_2)}{\partial K} \\ &= \frac{\partial E(y_1'y_1-2y_1'Ky_2 + y_2'K'Ky_2)}{\partial K} \end{aligned} \tag{2.31}$$

Since the required regularity conditions are satisfied the operators ∂ and E can be interchanged. Hence, we have

$$\begin{aligned} 0 &= -2E\frac{\partial y_1'Ky_2}{\partial K} + E\frac{\partial y_2'K'Ky_2}{\partial K} \\ &= -2Ey_1y_2' + 2KEy_2y_2' \\ &= -2\Gamma_{12} + 2K\Gamma_{22} \end{aligned} \tag{2.32}$$

which equation is identical with (2.22).

A last remark to be made concerns the minimum error variance property of the estimator Ky_2 even if the class of estimators is no longer confined to the class of linear estimators. This follows straightforwardly from the normality assumption of y_1 and y_2 and the well-known formula for (conditional) variances of two stochastic variables a and b

$$\begin{aligned} \text{var}(a) &= Ea^2 - (Ea)^2 \\ &= E\{E(a^2|b)\} - E\{E(a|b)\}^2 + E\{E(a|b)\}^2 - \{EE(a|b)\}^2 \\ &= E\{\text{var}(a|b)\} + \text{var}\{E(a|b)\} \end{aligned} \tag{2.33}$$

Let now f be an arbitrary vector function of y_2 of appropriate order then generalizing (2.33) for the multivariate case we obtain

$$\text{cov}\{y_1-f(y_2)\} = E\{\text{cov}(y_1-f(y_2)|y_2)\} + \text{cov}\{E(y_1-f(y_2)|y_2)\} \tag{2.34}$$

Taking traces of both sides of (2.34) we have in view of (2.29)

$$\begin{aligned} \text{tr}\{\text{cov}(y_1-f(y_2))\} &\geqq \text{tr}\{E[\text{cov}(y_1|y_2)]\} \;+\; 0 \\ &\geqq \text{tr}(\Gamma_{11}-\Gamma_{12}\Gamma_{22}^{-1}\Gamma_{21}) \end{aligned} \tag{2.35}$$

The equality sign applies when $f(y_2)=E(y_1|y_2)$. Hence, if normality is assumed $E(y_1|y_2)=Ky_2$ is the minimum variance estimator of y_1.

So the important conclusion we may draw from this subsection is the following. Let y_1 and y_2 be two multivariate normal vectors with zero mean and covariance matrices Γ_{11} and Γ_{22}, respectively, and $Ey_1y_2' = \Gamma_{12}$, then the best estimator of y_1 given y_2 is $E(y_1|y_2)=Ky_2$, where the Kalman matrix is given by $K=\Gamma_{12}\Gamma_{22}^{-1}$ whereas the error covariance matrix of this estimator is given by $\Gamma_{11}-\Gamma_{12}\Gamma_{22}^{-1}\Gamma_{21}$. In the next section we will show that this conclusion and its underlying geometric argument play an important role in the derivation of the Kalman filter equations. It should be noted once more that the conclusion drawn here and its geometric interpretation are well known (see, e.g., Åström (1970, pp. 222-224)). However, the derivations of K and Γ_u as given by (2.23) and (2.26) are, to my best knowledge, not mentioned elsewhere in the literature.

4.3 The Kalman filter equations

4.3.1 The state space model

Let y_t be a vector of m measurements or observations in period t and let y_t be related to a vector b_t of h unobserved components. The so called state vector b_t describes the state of the system in period t and is assumed to follow a vector AR(1) process. Such a system can be written as a state space model consisting of two equations

$$y_t = H_t b_t + \varepsilon_t \text{ (measurement equation)} \tag{3.1}$$

$$b_t = Ab_{t-1} + a_t \text{ (system equation)} \tag{3.2}$$

The aim of the Kalman filter equations is to evaluate (filter out) the unobserved components of b_t. H_t and A are known matrices of appropriate order. As regards the vectors of disturbances ε_t and a_t it is assumed that they are mutually independent and normally distributed with

$$E\varepsilon_t\varepsilon_s' = \delta_{ts}\sigma^2 R, \qquad E\varepsilon_t = 0 \tag{3.3}$$

$$Ea_t a_s' = \delta_{ts}\sigma^2 Q, \qquad Ea_t = 0 \tag{3.4}$$

$$E\varepsilon_t a_s' = 0 \quad \text{for every t and s} \tag{3.5}$$

where δ_{ts} stands for the Kronecker delta. Usually ε_t is called the measurement noise whereas a_t is referred to as the system noise.

The state space model can be employed for estimating and testing many types of models. To get an impression two examples are given.

Example 1 - A regression model with time varying parameters can be written as a state space model

$$y_t = x_t' b_t + \varepsilon_t \tag{3.6}$$

$$b_t = A b_{t-1} + a_t \tag{3.7}$$

In (3.6) y_t is a scalar, x_t' is a 1xk vector of k explanatory variables $x_t'=(x_{1t},\ldots,x_{kt})$, and $E\varepsilon_t^2=\sigma^2$; note that $H_t=x_t'$. The coefficients b_{jt} stacked in vector b_t are assumed to vary over time. Often a random walk is postulated, that is

$$b_t = b_{t-1} + a_t \tag{3.8}$$

where $Ea_t a_t'=\sigma^2 Q$. A practical problem is the determination of Q. Anyhow, the state space model here is a tool to verify the assumption whether the parameters of a regression model are constant or not. If one takes Q=0 which implies $b_t=b_{t-1}=b$, then the recursive Kalman filter equations for model (3.6)-(3.8) reduce to the well-known formula of the recursive least squares method in regression analysis; see Harvey (1981, pp. 192-193).

Example 2 - Any ARMA(p,q) process can be formulated as a state space model. To illustrate this consider the ARMA(2,1) process

$$y_t = \varphi_1 y_{t-1} + \varphi_2 y_{t-2} + \varepsilon_t - \lambda\varepsilon_{t-1} \tag{3.9}$$

where $E\varepsilon_t^2=\sigma^2$. Its representation in state space model form is

$$y_t = (1,0)\begin{bmatrix} b_{t,1} \\ b_{t,2} \end{bmatrix} \tag{3.10}$$

$$\begin{bmatrix} b_{t,1} \\ b_{t,2} \end{bmatrix} = \begin{bmatrix} \varphi_1 & 1 \\ \varphi_2 & 0 \end{bmatrix} \begin{bmatrix} b_{t-1,1} \\ b_{t-1,2} \end{bmatrix} + \begin{bmatrix} 1 \\ -\lambda \end{bmatrix} \varepsilon_t \tag{3.11}$$

where ε_t denotes the white noise error term in equation (3.9). The original

ARMA(2,1) process can be recovered from (3.10) and (3.11) by noting that $y_t=b_{t,1}$ and by substitution of the lower row into the upper row of (3.11). In this example there is no measurement noise.

4.3.2 A general geometric derivation of the Kalman filter equations

In the estimation procedure of the state space model the estimator $b_{t|s}$ plays an important role. The estimator $b_{t|s}$ is linear in the observations $y_1,\ldots,y_s$ and is defined as the best linear unbiased estimator of b_t given $y_1,\ldots,y_s$ and given a prior initial estimate $b_{0|0}$ of b_0 with error covariance matrix $\sigma^2 P_{0|0}$. The error covariance matrix $\sigma^2 P_{t|s}$ is defined by

$$\sigma^2 P_{t|s} = E(b_t - b_{t|s})(b_t - b_{t|s})' \tag{3.12}$$

The estimator $b_{t|s}$ is best in the least squares sense, i.e., $E\{b_t(r)-b_{t|s}(r)\}^2 =\|b_t(r)-b_{t|s}(r)\|^2$ is minimum within the class of linear unbiased estimators, $r=1,2,\ldots,h$. It should be emphasized that "best" in this sense also implies that the error covariance matrix of any other linear unbiased estimator exceeds $\sigma^2 P_{t|s}$ by a positive semidefinite matrix since $tr\{P_{t|s}\}$ is minimum as easily can be shown; see Magnus and Neudecker (1988, pp. 255-256). It should also be noted that "unbiased" is used in the sense that $E\{b_t-b_{t|s}\}=0$. The evaluation of $b_{t|t}$ is called filtering. In the literature $b_{t|s}$ is often called the linear least squares estimate (LLSE) of b_t; see Kailath (1981, p. 79). Given the initial values $b_{0|0}$ and $P_{0|0}$ the estimation can be carried out in a recursive manner for successively $t=1,2,\ldots,n$. The recursive formulae were originally proposed by Kalman (1960).

For every t two steps can be distinguished. First we have the prediction stage. Using induction we assume that $b_{t-1|t-1}$ is the best linear unbiased estimator of b_{t-1} given all information up to period t-1. Then equation (3.2) suggests the following estimator for b_t

$$b_{t|t-1} = Ab_{t-1|t-1} \tag{3.13}$$

It can easily be shown that this estimator is the best linear unbiased estimator of b_t given all information up to period t-1; see, for instance, Bertsekas (1976, p. 163). The error covariance matrix can directly be found by subtracting (3.13) from (3.2), yielding

$$b_t - b_{t|t-1} = A(b_{t-1} - b_{t-1|t-1}) + a_t \tag{3.14}$$

Taking variances of both sides of (3.14) and neglecting the scaling factor σ^2 we get

$$P_{t|t-1} = AP_{t-1|t-1}A' + Q \tag{3.15}$$

where use has been made of $cov(b_{t-1}-b_{t-1|t-1}, a_t)=0$.

Secondly, we have the update stage. This starts with the evaluation of the one step ahead prediction error

$$e_t = y_t - y_{t|t-1} = y_t - H_t b_{t|t-1} = H_t(b_t - b_{t|t-1}) + \varepsilon_t \tag{3.16}$$

which obviously has a zero mean. The sequence of the e_t is referred to as the innovations and it can be shown that they are mutually uncorrelated; we will show this in Section 4. The evaluation of $b_{t|t}$ can be carried out in a convenient manner by writing $b_{t|t}$ as

$$b_{t|t} = b_{t|t-1} + K_t e_t \tag{3.17}$$

where K_t is an h×m matrix. In order to justify the form of equation (3.17) define the stochastic error variable $\Delta b_{t|s} = b_t - b_{t|s}$ and note that the linear space spanned by the elements in $b_{0|0}, y_1, \ldots, y_t$ is identical with the space spanned by the elements in $b_{0|0}, e_1, \ldots, e_t$. Then consider the regression of the estimation error $\Delta b_{t|0} (= b_t - A^t b_{0|0})$ on $e_1, \ldots, e_t$ with regression matrices K_{tj}

$$\Delta b_{t|0} = \sum_{j=1}^{t} K_{tj} e_j + \Delta b_{t|t} \left(= \left[K_{t1} | \ldots | K_{tt}\right] \begin{bmatrix} e_1 \\ \vdots \\ e_t \end{bmatrix} + \Delta b_{t|t}\right) \tag{3.18}$$

where $\Delta b_{t|t}$ stands for the orthogonal and "unexplained" remainder of the regression equation; cf. (2.20). As we have seen in the previous section $E\{\Delta b_{t|t}(r)\}^2$ can be minimized for r=1,2,...,h by solving the corresponding t×h×m normal equations, i.e., postmultiply both sides of (3.18) by e_j' and take expectations, yielding

$$K_{tj} = E\{\Delta b_{t|0} e_j'\}[E e_j e_j']^{-1}, \quad j=1,\ldots,t \tag{3.19}$$

where use has been made of the zero covariance of e_i and e_j, $i \neq j$; cf. (2.23). In practice it will usually be hard to apply (3.19). In order to avoid (3.19) we write similar to (3.18)

$$\Delta b_{t|0} = \sum_{j=1}^{t-1} K_{tj} e_j + \Delta b_{t|t-1} \tag{3.20}$$

Subtracting (3.20) from (3.18) we obtain

$$\Delta b_{t|t-1} = K_{tt} e_t + \Delta b_{t|t} \tag{3.21}$$

or, by definition,

$$(b_t - b_{t|t-1}) = K_t e_t + (b_t - b_{t|t}) \tag{3.22}$$

which equation is identical to (3.17); note that matrix K_{tt} is denoted by K_t.

It is recalled that (3.22) has the same geometric interpretation as (2.20). The regression equation (3.22) is the key equation in the derivation of the formula for the so called Kalman gain K_t. It gives insight in the structure of the Kalman equations and it also explains that K_t can be interpreted as a matrix of regression coefficients. Peculiarly, equation (3.22) and its specific geometric interpretation have, to my best knowledge, not been mentioned elsewhere in the literature. As we have seen in the previous section we can obtain an expression for K_t in (3.22) by minimizing

$$\mathrm{tr}\, E(b_t - b_{t|t})(b_t - b_{t|t})' \tag{3.23}$$

or, equivalently, by solving the normal equations $E(b_t - b_{t|t})e_t' = 0$; cf. (2.21). Combining the latter equality and (3.22) we obtain

$$E(b_t - b_{t|t-1})e_t' = K_t\, Ee_t e_t' \tag{3.24a}$$

or

$$K_t = \{E(b_t - b_{t|t-1})e_t'\}(Ee_t e_t')^{-1} \tag{3.24b}$$

where the innovation e_t is defined by (3.16). Hence, two terms have to be evaluated. Using (3.16) we find for the first term

$$\begin{aligned} E(b_t - b_{t|t-1})e_t' &= E(b_t - b_{t|t-1})(y_t - H_t b_{t|t-1})' \\ &= E(b_t - b_{t|t-1})[H_t(b_t - b_{t|t-1}) + \varepsilon_t]' \\ &= \sigma^2 P_{t|t-1} H_t' \end{aligned} \tag{3.25}$$

and since

$$\begin{aligned} e_t &= y_t - H_t b_{t|t-1} \\ &= H_t(b_t - b_{t|t-1}) + \varepsilon_t \end{aligned} \tag{3.26}$$

we obtain for the second term

$$Ee_te_t' = \sigma^2(H_tP_{t|t-1}H_t' + R) \tag{3.27}$$

Combining (3.24b), (3.25) and (3.27) yields

$$K_t = P_{t|t-1}H_t'(H_tP_{t|t-1}H_t' + R)^{-1} \tag{3.28}$$

It is recalled from the previous section that by this choice of K_t every element in the orthogonal remainder $b_t-b_{t|t}$ in (3.22) is orthogonal to and stochastically independent of all elements in the regressor e_t, i.e., $E\Delta b_{t|t}e_t' = 0$, cf. (2.21). Hence, similar to (2.25) we may take the variances of both sides of (3.22) yielding

$$\sigma^2P_{t|t-1} = K_t(Ee_te_t')K_t' + \sigma^2P_{t|t} \tag{3.29}$$

From (3.24b) and (3.25) it can be seen that

$$K_t(Ee_te_t') = \sigma^2P_{t|t-1}H_t' \tag{3.30}$$

Now $P_{t|t}$ can be determined from (3.29) and (3.30), viz.

$$\begin{aligned} P_{t|t} &= P_{t|t-1} - P_{t|t-1}H_t'K_t' \\ &= P_{t|t-1} - K_tH_tP_{t|t-1} \\ &= (I - K_tH_t)P_{t|t-1} \end{aligned} \tag{3.31}$$

Summarizing and rewriting the foregoing equations (3.13), (3.15), (3.17), (3.28) and (3.31) we obtain

$$b_{t|t-1} = Ab_{t-1|t-1} \tag{3.32a}$$

$$P_{t|t-1} = AP_{t-1|t-1}A' + Q \tag{3.32b}$$

$$b_{t|t} = b_{t|t-1} + K_t(y_t - H_t b_{t|t-1}) \tag{3.32c}$$

$$P_{t|t} = (I - K_t H_t)P_{t|t-1} \tag{3.32d}$$

$$K_t = P_{t|t-1}H_t'(H_t P_{t|t-1}H_t' + R)^{-1} \tag{3.32e}$$

These five equations are referred to as the Kalman filter. For the sake of completeness we mention that combining (3.32b), (3.32d) and (3.32e) yields the following recursion for $P_{t+1|t}$

$$P_{t+1|t} = A\,[I - P_{t|t-1}H_t'(H_t P_{t|t-1}H_t'+R)^{-1}H_t]\;P_{t|t-1}A' + Q \tag{3.32f}$$

which is known as the Riccati equation.

It should be noted that the scaling factor σ^2 does not play a role in the Kalman filter. Furthermore, it follows from (3.3) that one should take R=1 in the case that y_t is a scalar variable unless $var(\varepsilon_t)=0$, then we have R=0; also note that in the scalar case only the variance ratio $var(a_t)/var(\varepsilon_t)=Q$ plays a role for determining K_t as can be seen from (3.32). Besides, it is worthwhile to notice that, if A, H_t, Q and R are known, $P_{t|t-1}$, $P_{t|t}$ as well as the Kalman gain K_t can be calculated in advance given $P_{0|0}$, even before data become available.

4.3.3 Comparison with other derivations

In this subsection we will give a brief survey of the main differences between the derivation of the Kalman filter equations given in the previous subsection and some standard derivations in the literature.

Firstly, we pay some attention to the derivation given by Harvey (1981). His derivation is strongly based on the derivation given by Duncan and

Horn (1972). In that approach the generalized least squares (GLS) principle is applied. However, as Harvey pointed out, in the derivation it is implicitly assumed that the matrix R is positive definite. Then referring to certain results from Theil (1971, pp. 282-287) Harvey states that for the univariate case the Kalman filter equations are still valid if R=0. In the derivation given here it does not make any difference whether R is singular or not.

In order to shed some light on the main difference between our geometric derivation and a number of standard derivations in the literature on systems theory it is useful to consider an interpretation of the one step ahead predictor $b_{t|t-1}$ in terms of conditional expectations. According to (2.28a), (3.19) and (3.20) we have

$$\begin{aligned} \Delta b_{t|0} &= \sum_{j=1}^{t-1} K_{tj} e_j + \Delta b_{t|t-1}, \qquad K_{tj} = E\{\Delta b_{t|0} e_j'\}[E e_j e_j']^{-1} \\ &= \sum_{j=1}^{t-1} E\{\Delta b_{t|0} | e_j\} + \Delta b_{t|t-1} \\ &= E(\Delta b_{t|0} | e_1, \ldots, e_{t-1}) + \Delta b_{t|t-1} \end{aligned} \tag{3.33}$$

It is recalled from the previous subsection that the expression for K_{tj} directly follows from the normal equations $E\Delta b_{t|t-1} e_j' = 0$ and the zero covariance of e_i and e_j $(i \neq j)$; note indeed that this zero covariance in combination with (2.28a) also directly imply the last equality of (3.33). Now using the definition of $\Delta b_{t|0}$ and $\Delta b_{t|t-1}$ it can be seen from (3.33) that

$$b_{t|t-1} = b_{t|0} + E(b_t - b_{t|0} | e_1, \ldots, e_{t-1}) \tag{3.34}$$

This interpretation can be applied to a broad class of state space models, i.e., the initial estimate of b_t, $b_{t|0} (= A^t b_{0|0})$, can be fixed or random. If $b_{0|0}$ and, consequently, $b_{t|0}$ are fixed then (3.34) becomes

$$
\begin{aligned}
b_{t|t-1} &= b_{t|0} + E(b_t|e_1,\ldots,e_{t-1}) - E(b_{t|0}|e_1,\ldots,e_{t-1}) \\
&= E(b_t|e_1,\ldots,e_{t-1}) \\
&= E(b_t|y_1,\ldots,y_{t-1}) \qquad (3.35)
\end{aligned}
$$

where the last equality follows from the fact that apart from a constant every element in y_j can be written as a linear combination of the elements in $e_1,\ldots,e_j$ $(j=1,\ldots,t-1)$. In other words, the elements in $e_1,\ldots,e_{t-1}$ contain the same information as the elements in $y_1,\ldots,y_{t-1}$, at least if $b_{0|0}(=Eb_0)$ is known and fixed. The conditional expectation (3.35) is often given as definition or interpretation in the literature; see Anderson and Moore (1979, p. 106), Åström (1970, p. 226), Chow (1983, p. 322), Goodwin and Sin (1984, p. 250) and Kalman (1960). However, one should be aware of the underlying assumption that $b_{0|0}(=Eb_0)$ is fixed and known. In the case that $b_{0|0}$ is random one should consider the slightly more general conditional expectation (3.34). This little difference becomes essential for a simple regression model as

$$
y_t = b + \varepsilon_t, \quad var(\varepsilon_t) = 1 \qquad (3.36a)
$$

or in state space form

$$
y_t = b_t + \varepsilon_t \qquad (3.36b)
$$

$$
b_t = b_{t-1} \; (= b) \qquad (3.36c)
$$

Contrary to one of the standard assumptions in systems theory we now assume $b(=Eb)$ is a completely unknown and fixed parameter to be estimated from the data. Provided $b_{0|0}=y_0$ and $P_{0|0}=E(b-y_0)^2=1$ it can easily be verified that the result of the Kalman filter equations (3.32) is identical to the well-known maximum likelihood estimator of b, that is

$$b_{t|t} = b_{t|t-1} + e_t/(t+1) = (y_0+\ldots+y_t)/(t+1) \tag{3.37a}$$

$$b_{t|t-1} = (y_0+\ldots+y_{t-1})/t, \quad e_t = y_t - b_{t|t-1} \tag{3.37b}$$

$$P_{t|t-1} = 1/t, \quad K_t = P_{t|t} = 1/(t+1) \tag{3.37c}$$

For the specific model (3.36b) and (3.36c) with var(b)=0 we have as a matter of fact

$$E(b_t|y_1,\ldots,y_{t-1}) = b \neq b_{t|t-1} \tag{3.38}$$

since b is a fixed parameter. According to the standard assumptions of a state space model in systems theory $b_{t|t-1}=E(b|y_1,\ldots,y_{t-1})$ would have been correct if b had been drawn in period 0 from a normal distribution with known mean $\bar{b}_0$=Eb and var(b)=1, then we could have set $b_{0|0}=\bar{b}_0$ and $P_{0|0}=1$. However, more generally, if $\bar{b}_0(=Eb_0)$ is an unknown parameter to be estimated from the data then $b_{0|0}$ is random rather than fixed. In practice this may occur fairly often and, consequently, the conditional expectation (3.34) should be adopted rather than (3.35). An alternative approach in such a case is to set $b_{0|0}=q$ and $P_{0|0}=mI_h$ with q being an arbitrary fixed vector and m being a large number reflecting our uncertainty; see Harvey and Phillips (1979). However, applying the latter approach the innovations e_t do have, in general, no longer zero expectation, although the bias becomes negligible if m is large enough.

Another interesting difference between several derivations in the literature and the derivation given in subsection (4.3.2) are three equalities which are often used in systems theory to derive the celebrated Kalman filter equations

$$E(b_t-b_{t|t-1})y_j' = 0, \quad j<t \tag{3.39a}$$

$$E(b_t-b_{t|t-1})b_{t|t-1}' = 0 \tag{3.39b}$$

$$Ee_ty_j' = 0, \quad j<t \tag{3.39c}$$

The first equality can be found in Anderson and Moore (1979, p. 106), Chow (1983, p. 327), Goodwin and Sin (1984, p. 250), Kailath (1981, p. 78) and Kalman (1960). The underlying argument is that the elements in $b_t - b_{t|t-1}$ are orthogonal to the space spanned by the elements in $y_1, \ldots, y_{t-1}$, since the elements in $b_{t|t-1}$ are the orthogonal projections of the elements in b_t onto this space. This also explains the second equality which can be found in Bertsekas (1976, p. 169), Åström (1970, p. 227) and Anderson and Moore (1979, p. 106). Their notation is somewhat different from ours. They denote $b_{t|t-1}$ by $\hat{x}(t)$ and $b_t - b_{t|t-1}$ by $\tilde{x}(t)$. Further, since $e_t = y_t - y_{t|t-1} = H_t(b_t - b_{t|t-1}) + \varepsilon_t$ (3.39c) follows from (3.39a). The third equality can be found in Åström (1970, p. 227), Goodwin and Sin (1984, p. 250), Hannan and Deistler (1988, p. 92), Kailath (1981, p. 77) and Kalman (1960).

In the sequel it is explained that the three equalities (3.39) are an immediate consequence of the standard assumption in systems theory that $b_{0|0}(=Eb_0)$ is fixed. If we drop this assumption the equalities do no longer need to hold. In order to illustrate this we consider again the simple regression model (3.36) in state space form

$$y_t = b_t + \varepsilon_t, \quad \mathrm{var}(\varepsilon_t) = 1 \tag{3.40a}$$

$$b_t = b_{t-1} \ (= b) \tag{3.40b}$$

Now given the random variable y_0 and its variance we obviously have

$$b_{0|0} = y_0, \quad P_{0|0} = E(y_0 - b)^2 = 1, \qquad b_{1|0} = b_{0|0} = y_0 \tag{3.41a}$$

$$P_{1|0} = 1, \quad K_1 = P_{1|0}/(P_{1|0}+1) = 1/2, \quad e_1 = y_1 - y_0 \tag{3.41b}$$

$$b_{2|1} = b_{1|1} = b_{1|0} + K_1 e_1 = y_0 + (y_1 - y_0)/2 = (y_0 + y_1)/2 \tag{3.41c}$$

where use has been made of (3.32). Now we have successively

$$E(b_2-b_{2|1})y_1 = E\left[b - \frac{y_0+y_1}{2}\right]y_1$$
$$= (b^2 - \tfrac{1}{2}b^2) - \tfrac{1}{2}Ey_1^2$$
$$= \tfrac{1}{2}b^2 - \tfrac{1}{2}(1+b^2) = -1/2 \neq 0 \qquad (3.42a)$$

$$E(b_2-b_{2|1})b_{2|1} = E(b-b_{2|1})(b_{2|1}-b+b)$$
$$= -E(b-b_{2|1})^2$$
$$= -\mathrm{var}\{(y_0+y_1)/2\} = -1/2 \neq 0 \qquad (3.42b)$$

and

$$Ee_2y_1 = E(b_2+\varepsilon_2-b_{2|1})y_1$$
$$= E(b-b_{2|1})y_1 = -1/2 \neq 0 \qquad (3.42c)$$

where (3.42c) follows from (3.42a). Furthermore, it should be stressed here that we also have $E(b_2-b_{2|1})y_1\neq 0$ even if $b_t=b_{t-1}+a_t$ with $\mathrm{var}(a_t)=Q>0$, $b_{0|0}=y_0$ and $P_{0|0}=1$. In that case only some more lines of straightforward algebra are needed. Similar remarks apply to (3.42b) and (3.42c).

These maybe somewhat counterintuitive results can be clarified by noting that the linear vector space spanned by the elements in $b_{0|0},y_1,\ldots,y_t$ is identical with the space spanned by the elements in $b_{0|0},e_1,\ldots,e_t$. Hence, assuming $E(b_0-b_{0|0})=0$ y_j can be written as a linear combination of the elements in $b_{0|0},e_1,\ldots,e_j$

$$y_j = H_jA^jb_{0|0} + M_{j1}e_1 + \ldots + M_{jj}e_j, \quad j=1,\ldots,t \qquad (3.43)$$

where use has been made of (3.2); note that $Ey_j=H_jA^jEb_0$. Using again $Ee_ie_j'=0$ $(i\neq j)$ the following expression can be established

$$Ey_je_t' = H_jA^jEb_{0|0}e_t', \qquad j<t \qquad (3.44)$$

Furthermore, if $b_{0|0}$ is assumed to be fixed, i.e., $\mathrm{cov}(b_{0|0})=0$, it follows from (3.44) that

$$Ey_j e_t' = H_j A^j b_{0|0} Ee_t' = 0, \quad j<t \tag{3.45}$$

By a similar argument it can also be shown that $E(b_t-b_{t|t-1})b_{t|t-1}'$ as well as $E(b_t-b_{t|t-1})y_j'$ for $j<t$ are zero if $b_{0|0}$ is fixed. Hence, the assumption that $b_{0|0}$ is fixed plays a crucial role in the derivations of the authors mentioned above, whereas in our derivation $Eb_{0|0}e_t'$ does not necessarily need to be zero. In the rather general derivation given in the previous subsection the elements in $b_{0|0}$ may be regarded as fixed or random. Note that in the case that $b_{0|0}$ is random with error covariance matrix $P_{0|0}(b)$ the derivations of the Kalman filter of the authors mentioned above, can be employed as well provided we redefine $z_t=y_t-H_t b_{t|0}$ and $x_t=b_t-b_{t|0}$ with $b_{t|0}=A^t b_{0|0}$. Then replacing y_t, b_t and b_{t-1} in (3.1) and (3.2) by z_t, x_t and x_{t-1}, respectively, we can start the Kalman recursions with $x_{0|0}=0$ and $P_{0|0}=P_{0|0}(b)$ and, consequently, we get $b_{t|s}=x_{t|s}+b_{t|0}$ (s=t-1,t).

Finally, we discuss two minor errors in the derivation of the Kalman equations given by Hannan and Deistler (1988, p. 93). They state that $E(b_t-b_{t|t-1})e_t'=0$ or in their terminology that e_t is orthogonal to $x(t)-\hat{x}(t|t-1)$. In order to shed some light on this mistake we consider the following counterexample according to the standard assumptions in systems theory

$$y_t = b_t + \varepsilon_t, \qquad \mathrm{var}(\varepsilon_t) = 1 \tag{3.46a}$$

$$b_t = b_{t-1} + a_t, \qquad \mathrm{var}(a_t) = Q \tag{3.46b}$$

$$b_0 \sim N(0,Q) \tag{3.46c}$$

So we obviously have

$$b_{0|0} = 0, \quad P_{0|0} = \mathrm{var}(b_0 - b_{0|0}) = Q, \quad b_{1|0} = 0 \tag{3.47a}$$

$$e_1 = y_1, \quad b_1 - b_{1|0} = b_0 + a_1 \tag{3.47b}$$

Hence, we obtain

$$\begin{aligned} E(b_1 - b_{1|0})e_1 &= E(b_0 + a_1)y_1 \\ &= E(b_0 + a_1)(b_1 + \varepsilon_1) \\ &= E(b_0 + a_1)(b_0 + a_1 + \varepsilon_1) \\ &= E(b_0 + a_1)^2 = 2Q \neq 0 \end{aligned} \tag{3.48}$$

From (3.20) and its geometric interpretation it can be seen that, in general, one can only say $E(b_t - b_{t|t-1})e_j' = 0$, $j<t$. Hannan and Deistler (op. cit., p. 93) also argue that $Ea_{t+1}e_t' = Eb_{t+1}e_t' = \sigma^2 AK_t P_{t|t-1}$ or in their notation "$E\{\xi(t)e(t)'\} = E\{x(t+1)e(t)'\} = K(t)\Sigma(t)$, because $\xi(t) \perp x(t)$". However, a_t is a white noise sequence and therefore, we obviously have $Ea_{t+1}e_t' = 0$; also note that $E(b_t - b_{t|t-1})e_t' = \sigma_t^2 P_{t|t-1}H_t'$ rather than zero as can be seen from (3.25). It emerges that the two minor errors in the derivation of Hannan and Deistler compensate each other and do not affect their final results which are correct. A more intuitive explanation that $E(b_t - b_{t|t-1})e_t'$ cannot be zero is as follows. The purpose of the recursive Kalman equations in the update stage is to improve the predictor $b_{t|t-1}$ in the least squares sense by using the new information contained in y_t and represented by e_t ($y_t = y_{t|t-1} + e_t$). Were $b_t - b_{t|t-1}$ and e_t uncorrelated then e_t would give no further information about b_t $b_{t|t-1}$ which means in fact $b_{t|t} = b_{t|t-1}$ or, more precisely, $b_t - b_{t|t} = b_t - b_{t|t-1}$. Moreover, from a mathematical point of view the zero covariance between b_t $b_{t|t-1}$ and e_t would imply $K_t = 0$ and, consequently, also $b_{t|t} = b_{t|t-1}$ as can be seen from (3.24b) and (3.17), respectively.

For expository purposes we once more consider the simple example (3.40) for t=0,1,2. Furthermore, we assume $b_{0|0}$ $(=y_0)$ to be random rather than fixed.

It is recalled from the classical regression analysis that

$$b_{2|2} = (y_0+y_1+y_2)/3 \tag{3.49}$$

In view of (3.18) consider first the regression equation of Kalman type

$$b-y_0 = K_{21}e_1 + K_{22}e_2 + (b-b_{2|2}) \tag{3.50a}$$

$$e_1 = y_1-y_0 \tag{3.50b}$$

$$e_2 = y_2 - (y_0+y_1)/2 \tag{3.50c}$$

Obviously e_1 and e_2 are one step ahead prediction errors. Then as in the classical regression analysis we consider a similar regression equation with an identical regressand but with different regressors

$$b-y_0 = k_{21}(y_1-y_0) + k_{22}(y_2-y_0) + (b-b_{2|2}) \tag{3.51}$$

Of course the regressors in (3.51) are simple linear transformations of the regressors in (3.50). Equation (3.51) is a multivariate regression equation as described in the previous section. Solving the normal equations (2.22) in their transposed form we obtain for the case at hand

$$\begin{bmatrix} k_{21} \\ k_{22} \end{bmatrix} = \begin{bmatrix} E(y_1-y_0)^2 & E(y_1-y_0)(y_2-y_0) \\ E(y_1-y_0)(y_2-y_0) & E(y_2-y_0)^2 \end{bmatrix}^{-1} \begin{bmatrix} E(b-y_0)(y_1-y_0) \\ E(b-y_0)(y_2-y_0) \end{bmatrix}$$

$$= \begin{bmatrix} 2 & 1 \\ 1 & 2 \end{bmatrix}^{-1} \begin{bmatrix} 1 \\ 1 \end{bmatrix} = 1/3 \begin{bmatrix} 2 & -1 \\ -1 & 2 \end{bmatrix} \begin{bmatrix} 1 \\ 1 \end{bmatrix} = \begin{bmatrix} 1/3 \\ 1/3 \end{bmatrix} \tag{3.52}$$

Upon substituting (3.52) into (3.51) and rearranging some terms we obtain again the well-known expression (3.49).

<u>Remark 1</u>. The orthogonal remainder $b-b_{2|2}$ in (3.51) is uncorrelated with y_1-y_0

and y_2-y_0 rather than with y_1 and y_2. Also it can be seen from (3.50a) that e_2 is orthogonal to and independent of the remainder $b-b_{2|2}$. More generally, we have $Ee_t(b_t-b_{t|t})'=0$. This equality may look somewhat counterintuitive since b_t seems to affect both $e_t=y_t-y_{t|t-1}$ and $b_t-b_{t|t}$. Anyhow this equality is quite useful for deriving in a straightforward manner an expression for the error covariance matrix $P_{t|t}$ as can be seen from (3.29)-(3.31).

Remark 2. By applying (3.32) and the results in (3.41) it can easily be verified that the recursive Kalman filter equations also lead to (3.49).

Remark 3. In the classical regression analysis the minimum variance property of (3.49) is proved by referring to the Cramér-Rao inequality whereas in the multivariate regression analysis or, equivalently, in the Kalman filter procedure the minimum variance property of $b-b_{2|2}$ in (3.51) is based on (2.35).

Remark 4. Equation (3.51) illustrates the Bayesian interpretation of Kalman filtering. The stochastic variable b as deviation from its prior mean y_0 is "explained" by the new information contained in y_1-y_0 and y_2-y_0, resulting in a posterior mean $b_{2|2}$.

An interesting conclusion we may draw from this subsection, is that one should be aware of some limitations of a number of standard derivations in systems theory, in particular if the derivation is applied to the classical regression model with unknown parameters. These standard derivations are based on the assumption that the initial estimate $b_{0|0}$(=Eb_0) is fixed and known whereas in the classical regression model Eb_0 is unknown and is to be estimated from the data. Hence, in such cases the estimate $b_{0|0}$ is random rather than fixed. The derivation given in subsection 4.3.2 is always valid irrespective of the fact that $b_{0|0}$ is fixed or random. This is not to criticize the standard derivations in systems theory, it only underlines the difference between the basic assumptions of the standard model in systems theory and the basic assumptions of the classical regression model.

Consequently, the use of the Kalman filter within the framework of a regression model, with constant or with time varying parameters, requires a slightly more general derivation, i.e., without the limiting assumptions that Eb_0 is known and that $b_{0|0}$ is fixed.

4.4 The likelihood function

Before deriving the log likelihood function of $y_1, y_2, \ldots, y_n$ it is recalled from multivariate analysis that the density function of vector a, say f(a), is equal to the density function of vector b, say g(b) if the Jacobian of the transformation a=Tb is unity. This can easily be shown

$$g(b) = f(a)\left|\frac{\partial a}{\partial b'}\right| = f(a)|T| = f(a) \tag{4.1}$$

This simple statement will appear to be effective for the derivation of the log likelihood function of the observations.

Now introducing $Y=(y_1,\ldots,y_n)$ as the m×n matrix of observations and $E=(e_1,\ldots,e_n)$ as the m×n matrix of the one step ahead prediction errors it can be seen from (3.16) that the Jacobian of the transformation is unity, that is

$$\left|\frac{\partial vec(E)}{\partial vec(Y)'}\right| = 1 \tag{4.2}$$

where vec(.) denotes that the columns of an m×n matrix are being stacked one on top of the other such that we obtain an mn×1 vector.

Next, we will show that the innovations e_t are uncorrelated. We can write e_{t+1} as

$$\begin{aligned} e_{t+1} &= y_{t+1} - H_{t+1} b_{t+1|t} \\ &= H_{t+1}(b_{t+1} - b_{t+1|t}) + \varepsilon_{t+1} \\ &= H_{t+1} A(b_t - b_{t|t}) + H_{t+1} a_{t+1} + \varepsilon_{t+1} \end{aligned} \tag{4.3}$$

It is recalled from (3.18) that the elements in the orthogonal remainder $b_t - b_{t|t}$ are orthogonal to the elements in the vector regressors $e_1,\ldots,e_t$. Hence, it follows from (4.3) and from the assumptions concerning the disturbances a_{t+1} and ε_{t+1} that e_{t+1} is uncorrelated with $e_1,\ldots,e_t$. Because we assumed ε_t and a_t to be Gaussian $e_1,\ldots,e_n$ are also stochastically independent.

Since the e_t are stochastically independent we can write down the log likelihood function of $y_1,\ldots,y_n$ in view of (4.1) and (4.2)

$$\ln g(y_1,\ldots,y_n; b_{0|0}, P_{0|0}) = \sum_{t=1}^{n} \ln f(e_t) \tag{4.4}$$

Then from (2.9) and (4.4) it follows that

$$\ln g(Y) = -\tfrac{1}{2} nm \ln(2\pi\sigma^2) - \tfrac{1}{2} \sum_{t=1}^{n} \ln|F_t| - \tfrac{1}{2} \sum_{t=1}^{n} (e_t' F_t^{-1} e_t)/\sigma^2 \tag{4.5}$$

where F_t is equal to

$$\begin{aligned} F_t &= (Ee_t e_t')/\sigma^2 \\ &= H_t P_{t/t-1} H_t' + R \end{aligned} \tag{4.6}$$

where use has been made of (3.27). Maximizing (4.5) with respect to σ^2 yields

$$0 = \frac{\partial \ln g(y)}{\partial \sigma^2}$$

$$= \sum_{t=1}^{n} \left[- \frac{m}{2\sigma^2} + \frac{e_t' F_t^{-1} e_t}{2\sigma^4} \right] \tag{4.7}$$

This implies

$$\tilde{\sigma}^2 = \frac{1}{nm} \sum_{t=1}^{n} e_t' F_t^{-1} e_t \tag{4.8}$$

Substitution of (4.8) into (4.5) yields the concentrated log likelihood function

$$\ln g_c(y) = - \frac{1}{2} nm \ln(2\pi\tilde{\sigma}^2) - \frac{1}{2} \sum_{t=1}^{n} \ln|F_t| - \frac{1}{2} nm \tag{4.9}$$

If we assume m=1 then (4.8) becomes

$$\tilde{\sigma}^2 = \frac{1}{n} \sum_{t=1}^{n} \frac{e_t^2}{f_t} \tag{4.10}$$

where the variance f_t of e_t/σ can be determined from (3.27)

$$f_t = H_t' P_{t|t-1} H_t + R \tag{4.11}$$

It is recalled that R=1 if y_t is a scalar and for the case that there is no measurement noise, we have R=0. Furthermore, the concentrated log likelihood function of the scalars $y_1, \ldots, y_n$ reduces to

$$\ln g_c(y_1, \ldots, y_n) = - \frac{1}{2} n \ln(2\pi\tilde{\sigma}^2) - \frac{1}{2} \sum_{t=1}^{n} \ln f_t - \frac{1}{2} n \tag{4.12}$$

where $\tilde{\sigma}_t^2$ and f_t are defined by (4.10) and (4.11), respectively. It should be noted that the log likelihood function given by (4.9) and (4.12) depends upon the initial values of $b_{0|0}$ and $P_{0|0}$. However, for the regression model with ARMA disturbances $b_{0|0}$ is usually unknown and is to be estimated from the first few observations. Therefore Harvey and Phillips propose an alternative method to calculate the exact likelihood function of all observations, which is different from the likelihood function obtained from the density function of the innovations e_t; see also Steyn (1988) for a further discussion. A survey of the method proposed by HP is given in the next two sections.

4.5 Estimation of linear models with ARMA(p,q) disturbances by means of Kalman filtering

Consider again the regression model from Chapter 2

$$y_t = x_t'\alpha + u_t \tag{5.1}$$

$$u_t = \varphi_1 u_{t-1} + \ldots + \varphi_p u_{t-p} + \varepsilon_t - \lambda_1 \varepsilon_{t-1} - \ldots - \lambda_q \varepsilon_{t-q} \tag{5.2}$$

where the ε_t are independent, and identically and normally distributed. Assuming stationarity Harvey and Phillips (1979) show how the exact likelihood function of $y_1,\ldots,y_n$ can be calculated given $x_1,\ldots,x_n$ and given the parameter set $\{\varphi,\lambda\}$ by employing the state space form of model (5.1) and (5.2) and by applying the Kalman filter technique.

To obtain the state space representation of (5.1) and (5.2) first the state space model formulation of (5.2) is given. Equation (5.2) can be written in state space form in various manners. One of these possibilities is

$$u_t = e_1'\delta_t \tag{5.3}$$

$$\delta_t = \begin{bmatrix} \varphi_1 & \vdots & \\ \varphi_2 & \vdots & I_{r-1} \\ \vdots & \vdots & \\ \cdots & \cdots & \cdots \\ \varphi_r & \vdots & 0' \end{bmatrix} \delta_{t-1} + \begin{bmatrix} 1 \\ -\lambda_1 \\ \vdots \\ -\lambda_{r-1} \end{bmatrix} \varepsilon_t \tag{5.4}$$

where e_1 is the first unit vector of I_r and r is defined by r=max(p,q+1); additional φ_i's and λ_i's are defined, as appropriate. Equation (5.2) can be recovered from (5.3) and (5.4) by successive substitutions starting at the bottom of the system. Note that u_t is equal to the first element of δ_t. Now the whole model (5.1) and (5.2) can be written in state space form

$$y_t = H_t b_t \tag{5.5}$$

$$b_t = \begin{bmatrix} \alpha_t \\ \\ \delta_t \end{bmatrix} = \begin{bmatrix} I & 0 \\ \\ 0 & T \end{bmatrix} \begin{bmatrix} \alpha_{t-1} \\ \\ \delta_{t-1} \end{bmatrix} + \begin{bmatrix} 0 \\ \\ w \end{bmatrix} \varepsilon_t \tag{5.6}$$

where H_t is a row vector of order k+r

$$H_t = (x_t', e_1')$$

whereas T and w are implicitly defined by (5.4). Note that R=0, i.e., there is no measurement noise.

The only thing which remains to be done before the recursive Kalman equations from the previous section can be applied is the determination of the initial values $b_{0|0}$ and $P_{0|0}$. For an estimate of α the first k observations are required. Applying the usual GLS formulae yields

$$\begin{aligned} \alpha_{k|k} &= (X_k'\Gamma_k^{-1}X_k)^{-1}X_k'\Gamma_k^{-1}y_k = X_k^{-1}y_k \\ &= X_k^{-1}X_k\alpha + X_k^{-1}u_k = \alpha + X_k^{-1}u_k \end{aligned} \tag{5.7}$$

where $X_k=(x_1,\dots,x_k)'$. Furthermore

$$E(\alpha_{k|k} - \alpha)(\alpha_{k|k} - \alpha)' = E\, X_k^{-1}u_k u_k'(X_k^{-1})'$$
$$= \sigma^2 X_k^{-1}\Gamma_k(X_k^{-1})' \tag{5.8}$$

The covariance matrix Γ_k of the first k disturbances can be determined from (7.8) in Chapter 2.

Assuming stationarity we have in view of (5.4) $E\delta_t=0$. Hence, the appropriate estimator for δ_k is a vector of zeroes

$$\delta_{k|k} = 0 \tag{5.9}$$

In order to determine the error covariance matrix of this estimator rewrite (5.4) as

$$\delta_t = T\delta_{t-1} + w\varepsilon_t \tag{5.10}$$

and taking the variances of both sides of (5.10) we obtain

$$D_0 = TD_0T' + \sigma^2 ww' \tag{5.11}$$

where $D_0 = E\delta_t\delta_t'$. Equation (5.11) can be solved by using the vec(.) operator. This yields

$$\mathrm{vec}(D_0) = (T\otimes T)\mathrm{vec}(D_0) + \sigma^2\mathrm{vec}(ww') \tag{5.12}$$

or

$$\mathrm{vec}(D_0) = \sigma^2(I-T\otimes T)^{-1}\mathrm{vec}(ww') \tag{5.13}$$

Hence, the error covariance matrix $\sigma^2 P_{k|k}$ of $b_{k|k}=[(X_k^{-1}y_k)',0']'$ is equal to

$$\sigma^2 P_{k|k} = \begin{bmatrix} \sigma^2 X_k^{-1}\Gamma_k(X_k^{-1})' & -E(\alpha_k-\alpha)\delta_k' \\ -E\delta_k(\alpha_k-\alpha)' & D_0 \end{bmatrix} \tag{5.14}$$

HP state that $E\delta_k(\alpha_k-\alpha)'$ can be evaluated in a relatively straightforward way for a pure moving average process. However, they remark, for general autoregressive-moving average processes it is more difficult to compute these matrices directly. They propose the following alternative approximation for MA as well as ARMA processes. Use all observations and set $b_{0|0}=0$ and

$$P_{0|0} = \begin{bmatrix} mI_k & 0 \\ 0 & \sigma^{-2}D_0 \end{bmatrix} \tag{5.15}$$

where m is a large scalar number reflecting the uncertainty about the true value of α. HP state that after k recursions the state vectors and associated error covariance matrices will be approximately equal to those obtained by the exact procedure in which $b_{k|k}$ and $P_{k|k}$ are computed directly. When applying the approximate method the first k prediction errors must be discarded as they depend on the arbitrary values chosen for the first k elements in $b_{0|0}$.

In their Monte Carlo study the calculations for the MA(1) case are carried out in single position and in double position and the best results are obtained for $m=10^6$ and $m=10^8$, respectively. Comparing the exact method and the approximate method in the case of MA(1) disturbances the authors conclude that the degree of approximation involved will be negligible.

It would be interesting to compare the exact method and the approximate method for general ARMA processes as well. Since a crucial difference between an MA process and an ARMA process is that the covariance matrix of the former process is always positive definite while the covariance matrix of the latter process can be nearly singular, particularly if one of the roots of the lag polynomial $\varphi(B)$ of the AR part of the process is close to the unit circle.

This might affect the results obtained by the application of the approximate equation (5.15). Moreover, the optimal choice of m seems to be machine dependent; see HP (op. cit.). Therefore, it also seems to be useful to derive an exact expression for $E\delta_k(\alpha_k-\alpha)'$ in the general case of an arbitrary stationary ARMA process. This can be done as follows.

From (5.2) and the first system equation at the top of (5.4) it appears that u_k can be written as

$$\delta_{k,1} = u_k = \varphi_1 u_{k-1} + \delta_{k-1,2} + \varepsilon_k \tag{5.16}$$

Next, it can be seen from (5.2) and (5.16) that

$$\delta_{k-1,2} = \varphi_2 u_{k-2} + \dots + \varphi_r u_{k-r} - \lambda_1 \varepsilon_{k-1} - \dots - \lambda_{r-1}\varepsilon_{k-r+1} \tag{5.17}$$

And, subsequently, in view of (5.17) and the second equation of (5.4) we obtain

$$\delta_{k-2,3} = \varphi_3 u_{k-3} + \dots + \varphi_r u_{k-r} - \lambda_2 \varepsilon_{k-2} - \dots - \lambda_{r-1}\varepsilon_{k-r+1} \tag{5.18}$$

Hence, by the principle of mathematical induction it can be shown

$$\delta_{k-i,1+i} = \sum_{j=1+i}^{r} \varphi_j u_{k-j} - \sum_{j=i}^{r-1} \lambda_j \varepsilon_{k-j}, \quad \lambda_0 = -1 \quad (i=0,1,\dots,r-1) \tag{5.19}$$

Premultiplying both sides of (5.19) by F^i, F being the forward shift operator ($Fx_t=x_{t+1}$), yields

$$\begin{aligned}\delta_{k,1+i} &= \sum_{j=1+i}^{r} \varphi_j u_{k+i-j} - \sum_{j=i}^{r-1} \lambda_j \varepsilon_{k+i-j} \\ &= \sum_{j=1+i}^{r} (\varphi_j u_{k+i-j} - \lambda_{j-1}\varepsilon_{k+i-j+1}), \quad i=0,1,\dots,r-1\end{aligned} \tag{5.20}$$

or more conveniently

$$\delta_{k,i} = \sum_{j=i}^{r} (\varphi_j u_{k+i-j-1} - \lambda_{j-1}\varepsilon_{k+i-j}), \quad i=1,2,\ldots,r \tag{5.21}$$

Furthermore, in view of (5.7) the hth element of $\alpha_{k|k}-\alpha$ can be written as

$$\alpha_{k|k}(h) - \alpha_h = \sum_{m=1}^{k} X_k^{-1}(h,m)\, u_m \tag{5.22}$$

Hence the elements of $C_k = E\delta_k(\alpha_{k|k}-\alpha)'$ can be determined from (5.21) and (5.22)

$$C_k(i,h) = \sum_{j=i}^{r} \sum_{m=1}^{k} X_k^{-1}(h,m)(\varphi_j \gamma_{m-k-i+j+1} - \lambda_{j-1}\nu_{m-k-i+j}) \tag{5.23}$$

where $\gamma_j = Eu_t u_{t-j}$ and $\nu_j = Eu_t \varepsilon_{t-j}$; γ_j and ν_j can be determined from the formulae (7.4), (7.8), (7.9) and (7.10) in Chapter 2.

Now the elements of C_k can be inserted in (5.14) and the recursive Kalman filter equations can be applied starting from an exact expression for $P_{k|k}$ for any stationary ARMA(p,q) process.

4.6 The exact likelihood function for models with ARMA(p,q) disturbances

Harvey and Phillips (1979) derive an algorithm for calculating the exact log likelihood function of the scalar observations $y_1, y_2, \ldots, y_n$. As shown in Section 2 the concentrated log likelihood function can be written as

$$\ln f(y) = -\tfrac{1}{2}n(1 + \ln \tfrac{2\pi}{n}) - \tfrac{1}{2}n \ln(Py-PX\tilde{\alpha})'(Py-PX\tilde{\alpha}) - \tfrac{1}{2}\ln|\Gamma| \tag{6.1}$$

where P is the lower triangular matrix such that $P'P = \Gamma^{-1}$ and $\tilde{\alpha}$ is the GLS estimator of α; cf (2.12). Note that the GLS estimator is equal to the ML estimator since it is assumed that all parameters of Γ are known. It emerges from (6.1) that only two terms on the right hand side of (6.1) are to be evaluated.

Since in the Kalman terminology $e_t = y_t - y_{t|t-1}$ can be seen as the one step ahead prediction error from the original regression model

$$y = X\alpha + u \tag{6.2}$$

$p_{tt}e_t$ can be seen as the one step ahead prediction error of the transformed regression model

$$Py = PX\alpha + Pu \tag{6.3a}$$

or

$$p_{tt}y_t = -\sum_{s=1}^{t-1} p_{ts}y_s + \sum_{s=1}^{t} p_{ts}\Big(\sum_{j=1}^{k} X_{js}\alpha_j + u_s\Big) \tag{6.3b}$$

And as model (6.3) obeys the classical assumptions it can be shown that the sum of the squared OLS residuals from (6.3a) is equal to the sum of the squared recursive residuals from (6.3a)

$$(Py-PX\tilde{\alpha})'(Py-PX\tilde{\alpha}) = \sum_{t=k+1}^{n} v_t^2 \tag{6.4}$$

where v_t denotes the recursive residual or standardized one step ahead prediction error from model (6.3a). Furthermore, v_t can be calculated from

$$\begin{aligned} v_t &= \sigma p_{tt}e_t/\{\mathrm{var}(p_{tt}e_t)\}^{1/2} \\ &= e_t/f_t^{1/2} \end{aligned} \tag{6.5}$$

where f_t can be determined from (4.11); note that R=0. For a proof of (6.4) see Phillips and Harvey (1974).

Now the only problem lies in the calculation of $|\Gamma|$. Consider the disturbance

$$u_t = \varphi_1 u_{t-1} + \dots + \varphi_p u_{t-p} + \varepsilon_t - \lambda_1 \varepsilon_{t-1} - \dots - \lambda_q \varepsilon_{t-q} \tag{6.6}$$

HP pointed out that the easiest way to calculate $|\Gamma|$ in the present context is to consider the state space representation (5.3) and (5.4) of the disturbances u_t. Since for this specific state space model $b_{0|0}=0$ and $P_{0|0}=\sigma^{-2}D_0$, D_0 being defined by (5.13), it is not a serious problem to start the recursive Kalman filter equations once again in order to obtain a set of n one step ahead prediction errors $\{e_{2t}\}$ which are independently, identically and normally distributed with zero expectation and variance $\sigma^2 f_{2t}$. Furthermore, the e_{2t} can be seen as a transformation of the u_t, the Jacobian of the transformation being unity. Thus if $e_2=Tu$ with $|T|=1$ then

$$\begin{aligned} |\Gamma| &= |\sigma^{-2}Euu'| \\ &= |\sigma^{-2}T^{-1}Ee_2e_2'(T^{-1})'| \\ &= |\sigma^{-2}Ee_2e_2'| \\ &= \prod_{t=1}^{n} f_{2t} \end{aligned} \tag{6.7}$$

where f_{2t} can be determined from (4.11) with R=0. Note that f_{2t} as well as the Kalman gain K_{2t} can be calculated in a recursive manner even without data. Only the parameter set $\{\varphi,\lambda\}$ is required.

Hence, in view of (6.4) and (6.7) the concentrated log likelihood function (6.1) becomes

$$\ln f(y) = -\tfrac{1}{2}n(1 + \ln \tfrac{2\pi}{n}) - \tfrac{1}{2}n \ln \sum_{t=k+1}^{n} v_t^2 - \tfrac{1}{2} \sum_{t=1}^{n} \ln f_{2t} \tag{6.8}$$

In order to obtain the ML estimates of the parameter set $\{\varphi,\lambda,\alpha,\sigma^2\}$ one has to maximize (6.8) with respect to φ and λ. The ML estimates of α and σ^2 are produced by the corresponding Kalman filter equations and (4.10). When the approximate equation (5.15) for $P_{0|0}$ is used HP advise to ignore the first k one step ahead prediction errors $e_1,\ldots,e_k$ since they depend on the arbitrary starting values chosen for the k elements in $\alpha_{0|0}$.

Their Monte Carlo experiments in the case that there is an MA(1) process among the disturbances bring them to conclude that loss of precision entailed by the use of the approximate method appears to be very small. Furthermore, they pointed out that by the employment of double precision or by the use of a square root algorithm the differences between the exact and approximate methods effectively disappear.

4.7 Predictions and prediction intervals by using Kalman filtering

Within the framework of the state space model and the Kalman filter it is easy to produce predictions and prediction intervals. For simplicity we consider the univariate case with $\mathrm{var}(\varepsilon_t)=\sigma^2$. The prediction made in period n for m periods ahead equals

$$y_{n+m|n} = H_{n+m} b_{n+m|n} \tag{7.1}$$

$$b_{n+m|n} = A^m b_{n|n} \tag{7.2}$$

while its error variance $V(y_{n+m|n})$ equals

$$V(y_{n+m|n}) = \{H_{n+m}[A^m P_{n|n}(A')^m + \sum_{j=0}^{m-1} A^j Q(A')^j]H'_{n+m} + 1\}\sigma^2 \tag{7.3}$$

See Abraham and Ledolter (1983). After replacing σ^2 by its ML estimate

$$\tilde{\sigma}^2 = \frac{1}{n-k} \sum_{t=k+1}^{n} \frac{(y_t - y_{t|t-1})^2}{H_t' P_{t|t-1} H_t + 1} \tag{7.4}$$

a prediction interval can be derived in the usual way by introducing the Student's t-distribution.

In practical situations it may occur that one is interested in the sum of m consecutive predictions separately, particularly in the case of a flow variable or when the time series under consideration consists of the first differences of the original time series. The error of such a sum equals

$$sy_{n+m|n} = \sum_{j=1}^{m} (y_{n+j} - y_{n+j|n}) \tag{7.5}$$

Define $V(sy_{n+m|n})$ as the error variance of $sy_{n+m|n}$ then it can be shown

$$V(sy_{n+m|n}) = (W_m P_{n|n} W_m' + Z_m + m)\sigma^2 \tag{7.6}$$

where

$$W_m = \sum_{j=1}^{m} H_{n+j} A^j, \quad W_0 = 0 \tag{7.7}$$

$$Z_m = \sum_{p=1}^{m} R_{p,m} Q R_{p,m}', \quad R_{p,m} = \sum_{j=p}^{m} H_{n+j} A^{j-p} \tag{7.8}$$

The proof of (7.6) follows from repeated substitution of (3.2) into (3.1)

$$\begin{aligned} y_{n+j} &= H_{n+j} b_{n+j} + \varepsilon_{n+j} \\ &= H_{n+j}(Ab_{n+j-1} + a_{n+j}) + \varepsilon_{n+j} \\ &= H_{n+j} A^j b_n + H_{n+j} \sum_{i=0}^{j-1} A^i a_{n+j-i} + \varepsilon_{n+j} \end{aligned} \tag{7.9}$$

Now making use of $y_{n+j|n} = H_{n+j} A^j b_{n|n}$ it follows from (7.9)

$$V(sy_{n+m|n}) = (W_m P_{n|n} W_m' + Z_m + m)\,\sigma^2 \tag{7.10}$$

where W_m is defined by (7.7) and Z_m stands for

$$\begin{aligned} Z_m &= \sigma^{-2} \operatorname{cov}\Big(\sum_{j=1}^{m} \sum_{i=0}^{j-1} H_{n+j} A^i a_{n+j-i} \Big) \\ &= \sigma^{-2} \operatorname{cov}\Big(\sum_{j=1}^{m} \sum_{p=1}^{j} H_{n+j} A^{j-p} a_{n+p} \Big) \\ &= \sigma^{-2} \operatorname{cov}\Big(\sum_{p=1}^{m} \sum_{j=p}^{m} H_{n+j} A^{j-p} a_{n+p} \Big) \\ &= \sum_{p=1}^{m} R_{p,m} Q R_{p,m}' \end{aligned} \tag{7.11}$$

Similar to the approach of Box and Jenkins (1976) these error variance formulae are only exact if the parameter set $\{\varphi,\lambda\}$ is known. The foregoing results can easily be generalized for the multivariate case. A method which allows for the uncertainty due to unknown parameters is given in Hamilton (1986).

CHAPTER V

ESTIMATION OF REGRESSION MODELS WITH MISSING OBSERVATIONS AND SERIALLY CORRELATED DISTURBANCES

5.1 Introduction

It may occur that disturbances in a regression model are serially correlated and that some observations are missing. Savin and White (1978a) and Richardson and White (1979) present various Durbin-Watson type tests for autocorrelation when observations are missing. Wansbeek and Kapteyn (1985) describe a maximum likelihood estimator and several two-step estimators of the model parameters for the case that the disturbances follow a first order Markov process. Their method is partly based on the well-known first order autocorrelation transformation of the data with a specific adjustment for the cases where data are missing.

The main aim of this chapter is to derive an appropriate transformation of the data of a regression model when the disturbances follow a second order autocorrelation process and when some observations are missing. A generalized Prais-Winsten two-step estimation procedure is given. Furthermore, attention is paid to a test procedure which discriminates between several types of autocorrelation. The results can easily be generalized for higher order autocorrelation processes.

The plan of this chapter is as follows. In Section 2 the model is described. In Section 3 an appropriate transformation of the data is presented. In Section 4 a two-step estimator is given and, additionally, attention is paid to the likelihood ratio test for second order autocorrelation when observations are missing. Furthermore, it is argued that the power of the Lagrange multiplier test can be improved through an adjustment for missing data. Finally, Section 5 is concerned with the Kalman equations which can also be applied in the case of missing data.

5.2 The model

The single equation regression model can be written in matrix notation as

$$y = X\alpha + u \tag{2.1}$$

where y and u are n×1 vectors, X is a n×k matrix of exogenous variables and α is the k×1 vector of parameters to be estimated. The elements u_t of u are assumed to follow a stationary second order autocorrelation (AR(2)) process

$$u_t = \varphi_1 u_{t-1} + \varphi_2 u_{t-2} + \varepsilon_t \tag{2.2}$$

where

$$\varphi_1 + \varphi_2 < 1 \tag{2.3a}$$

$$\varphi_2 - \varphi_1 < 1 \tag{2.3b}$$

$$\varphi_2 > -1 \tag{2.3c}$$

It is assumed that the ε_t are independent and identically distributed (iid) N(0,1). Most textbooks do not give the derivation of the stationarity conditions (2.3). Harvey (1981), for instance, refers to Goldberg (1958). For the sake of completeness a derivation is given in Appendix A of this chapter.

Next, we suppose that n-m observations are missing and that the actual sample size is m whereas the actual observations have the rank numbers $t_1, t_2, \ldots, t_m$ (t_1=1 and t_m=n). Deleting the missing observations in (2.1) we obtain (cf. Wansbeek and Kapteyn (1985))

$$y_* = X_*\alpha + u_* \tag{2.4}$$

Define $\Gamma_*=Eu_*u_*'$. Then after an appropriate m×m matrix P has been derived in such a way that $P\Gamma_*P'=I_m$, (2.4) can be transformed in

$$Py_* = PX_*\alpha + Pu_* \tag{2.5}$$

where $EPu_*u_*'P'=I_m$. Of course P depends on φ_1 and φ_2. When P is known, applying OLS to (2.5) gives the GLS estimator of α.

Before we derive an appropriate m×m matrix P we examine the covariances

$$\nu_j = Eu_t\varepsilon_{t-j} \quad \text{and} \quad \gamma_j = Eu_tu_{t-j} \tag{2.6}$$

It should be noted that $\nu_j=0$ for $j<0$ and it is easy to derive

$$\nu_0 = 1 \tag{2.7a}$$

$$\nu_1 = \varphi_1 \tag{2.7b}$$

$$\nu_j = \varphi_1\nu_{j-1} + \varphi_2\nu_{j-2}, \quad j=2,3,\ldots \tag{2.7c}$$

Upon multiplying both sides of (2.2) by u_{t-j} (j=0,1,2) and taking expectations we obtain

$$\gamma_0 = \varphi_1\gamma_1 + \varphi_2\gamma_2 + 1 \tag{2.8a}$$

$$\gamma_1 = \varphi_1\gamma_0 + \varphi_2\gamma_1 \tag{2.8b}$$

$$\gamma_2 = \varphi_1\gamma_1 + \varphi_2\gamma_0 \tag{2.8c}$$

From a similar operation for $j>2$ it follows that

$$\gamma_j = \varphi_1\gamma_{j-1} + \varphi_2\gamma_{j-2}, \quad j=3,4,\ldots \tag{2.9}$$

Solving (2.8) yields (see Judge et al. (1980, p. 190))

$$\gamma_0 = \frac{1-\varphi_2}{(1+\varphi_2)\{(1-\varphi_2)^2 - \varphi_1^2\}} \tag{2.10a}$$

$$\gamma_1 = \frac{\varphi_1}{1-\varphi_2}\gamma_0 \tag{2.10b}$$

$$\gamma_2 = \left[\varphi_2 + \frac{\varphi_1^2}{1-\varphi_2}\right]\gamma_0 \tag{2.10c}$$

It is well known that the lower triangular transformation matrix P of an AR(p) process is a band matrix with bandwidth p+1. This also holds for the case that the disturbances follow an AR(1) process while some observations are missing; see Wansbeek and Kapteyn (1985). However, in the case of AR(p) disturbances and missing observations the structure of the lower triangular transformation matrix does not necessarily have bandwidth p+1 as we will show in the sequel of this section. Consider the AR(2) process

$$u_t = u_{t-1} - .5u_{t-2} + \varepsilon_t \tag{2.11}$$

For the sake of simplicity we set $\gamma_0=1$. Then according to (2.9) and (2.10) we have

$$\gamma_0=1 \qquad \gamma_1=2/3 \qquad \gamma_2=1/6 \qquad \gamma_3=-1/6 \tag{2.12a}$$

$$\gamma_4=-1/4 \qquad \gamma_5=-1/6 \qquad \gamma_6=-1/24 \qquad \gamma_7=1/24 \tag{2.12b}$$

Next, we will give three examples. Firstly, define the column vector $\psi=(u_1,u_3,u_5,u_7)'$ and $\Gamma(\psi)$ as the corresponding covariance matrix of ψ and $P(\psi)$ as the corresponding lower triangular transformation matrix with $P(\psi)\Gamma(\psi)P(\psi)'=I_4$. Then $P(\psi)$ can be calculated by using the formulae (2.3.5), (2.3.7) and (2.3.9). This yields

$$P(\psi) = \begin{bmatrix} 1 & 0 & 0 & 0 \\ -0.1690 & 1.0142 & 0 & 0 \\ 0.3024 & -0.2268 & 1.0583 & 0 \\ -0.0707 & 0.3182 & -0.2475 & 1.0607 \end{bmatrix} \qquad (2.13)$$

Furthermore, define $\xi=(u_1,u_2,u_5,u_6)'$ then the corresponding transformation matrix $P(\xi)$ becomes

$$P(\xi) = \begin{bmatrix} 1 & 0 & 0 & 0 \\ -0.8944 & 1.3416 & 0 & 0 \\ 0.2582 & 0 & 1.0328 & 0 \\ -0.2309 & 0.3464 & -0.9238 & 1.3856 \end{bmatrix} \qquad (2.14)$$

Finally, consider the 3th example and define $\vartheta=(u_1,u_2,u_3,u_5,u_6)'$ then $P(\vartheta)$ is equal to

$$P(\vartheta) = \begin{bmatrix} 1 & 0 & 0 & 0 & 0 \\ -0.8944 & 1.3416 & 0 & 0 & 0 \\ 0.7746 & -1.5492 & 1.5492 & 0 & 0 \\ 0 & 0.5477 & -0.5477 & 1.0954 & 0 \\ 0 & -0.1826 & 0.5477 & -1.0954 & 1.4606 \end{bmatrix} \qquad (2.15)$$

These three examples illustrate that in the case of AR(2) disturbances and missing observations P does not always have bandwidth 3. Furthermore, example 3 suggests that the bandwidth depends on the pattern of missing observations, especially the position of two consecutive observations. In the following section we will show that this is indeed the case. It will turn out that the appropriate transformation of the disturbance u_t and, consequently, the transformations of the variables y_t and x_t in period t will never involve more than one preceding pair of two consecutive observations.

5.3 Derivation of the transformation matrix

We restrict ourselves to a derivation of the last row vector of the lower triangular m×m matrix P; the remaining m-1 rows of P can be derived in a similar way.

Let q be the largest number such that $q<m$ and $t_q - t_{q-1} = 1$. Further define $h = t_m - t_q$ and $r = m-q$. To clarify, the couple of observations with the rank numbers t_q and t_{q-1} is the nearest pair of two consecutive actual observations before observation $n = t_m$ and the distance between t_q and n is denoted by h. Hence, for the final h time periods the actual sample of r observations does not contain any consecutive observations except for the case where $r>1$ and the observations n-1 and n both happen to be available.

It follows from (2.2) by repeated substitution that u_n can be written as a linear combination of u_{n-h}, u_{n-h-1}, (i.e., u_{t_q} and $u_{t_{q-1}}$) and the innovations ε_{n-h+1} through ε_n in the following form

$$u_n = \sum_{j=0}^{h-1} \beta_j \varepsilon_{n-j} + \beta_{n,h} u_{n-h} + \beta_{n,h+1} u_{n-h-1} \tag{3.1}$$

From multiplying both sides of (3.1) by ε_{n-j} for $j=0,1,\ldots,h-1$ it can be seen that $\beta_j = \nu_j$ while it follows from multiplying by u_{n-h} and u_{n-h-1}, respectively, and by taking expectations that

$$\begin{aligned} \gamma_h &= \beta_{n,h}\gamma_0 + \beta_{n,h+1}\gamma_1 \\ \gamma_{h+1} &= \beta_{n,h}\gamma_1 + \beta_{n,h+1}\gamma_0 \end{aligned}$$

Hence,

$$\begin{bmatrix} \beta_{n,h} \\ \beta_{n,h+1} \end{bmatrix} = \begin{bmatrix} \gamma_0 & \gamma_1 \\ \gamma_1 & \gamma_0 \end{bmatrix}^{-1} \begin{bmatrix} \gamma_h \\ \gamma_{h+1} \end{bmatrix} \tag{3.2}$$

Similarly, the disturbances $u_{n-h+1},\ldots,u_{n-1}$ can be written as a linear combination of u_{n-h}, u_{n-h-1} and the innnovations. We find

$$u_t = \sum_{j=0}^{t-n+h-1} \nu_j \varepsilon_{t-j} + \beta_{t,h} u_{n-h} + \beta_{t,h+1} u_{n-h-1}, \quad t=n-h+1,\dots,n-1 \quad (3.3)$$

where

$$\begin{bmatrix} \beta_{t,h} \\ \beta_{t,h+1} \end{bmatrix} = \begin{bmatrix} \gamma_0 & \gamma_1 \\ \gamma_1 & \gamma_0 \end{bmatrix}^{-1} \begin{bmatrix} \gamma_{t-n+h} \\ \gamma_{t-n+h+1} \end{bmatrix} \quad (3.4)$$

If we now define $z_t = u_t - \beta_{t,h} u_{n-h} - \beta_{t,h+1} u_{n-h-1}$ for $t=n-h+1,\dots,n$ then it follows from (3.1) and (3.3) that

$$\begin{bmatrix} z_{n-h+1} \\ z_{n-h+2} \\ \vdots \\ z_n \end{bmatrix} = \begin{bmatrix} 1 & 0 & \dots & 0 \\ \nu_1 & 1 & \dots & 0 \\ \vdots & & & \vdots \\ & & & 0 \\ \nu_{h-1} & \dots & \nu_1 & 1 \end{bmatrix} \begin{bmatrix} \varepsilon_{n-h+1} \\ \varepsilon_{n-h+2} \\ \vdots \\ \varepsilon_n \end{bmatrix} \quad (3.5)$$

In the sequel of this section we will show that for the appropriate transformation of the data in period $n=t_m$ it suffices to take into account the disturbances and the data in the periods $t_{m-r-1}, t_{m-r}, \dots, t_m$ ($t_{m-r-1}=n-h-1$). It turns out that the appropriately transformed disturbance $\bar{u}_m$ can be written as a linear combination of $z_{n-h+1}, \dots, z_n$ and, consequently, in view of (3.5) as a combination of $\varepsilon_{n-h+1}, \dots, \varepsilon_n$. Therefore, $\bar{u}_m$ is uncorrelated with the previous disturbances $u_1, u_2, \dots, u_{n-h}$. Furthermore, it will be proved that $\bar{u}_m$ is also uncorrelated with the remaining disturbances of the available observations $u_{t_{q+1}}$ through $u_{t_{m-1}}$.

Define h_n as the projection of u_n onto the subspace S_{m-1} which is generated by $u_{n-h-1}, u_{n-h}, u_{t_{q+1}}, u_{t_{q+2}}, \dots, u_{t_{m-1}}$ and let w_n be the corresponding perpendicular or the orthogonal remainder, i.e.,

$$w_n = u_n - h_n \tag{3.6}$$

Then of course w_n is orthogonal to all elements of S_{m-1}. Hence,

$$Ew_n u_{t_{m-j}} = 0, \quad j=1,2,...,r+1 \tag{3.7}$$

See Section 3 of Chapter 2 for a definition of a linear vector space of stochastic variables.

Furthermore, w_n can be written as

$$w_n = u_n - \sum_{j=1}^{r+1} \psi_j u_{t_{m-j}} \tag{3.8}$$

The ψ_j follow from the solution of the set of r+1 normal equations $Ew_n u_{t_{m-j}}=0$, for j=1,...,r+1. Postmultiplying both sides of (3.8) by $u_{t_{m-j}}$ and taking expectations we obtain the so called least squares solution

$$\begin{bmatrix} \psi_1 \\ \vdots \\ \psi_{r+1} \end{bmatrix} = \begin{bmatrix} \zeta_{11} & \cdots & \zeta_{1,r+1} \\ \vdots & & \vdots \\ \zeta_{r+1,1} & \cdots & \zeta_{r+1,r+1} \end{bmatrix}^{-1} \begin{bmatrix} \zeta_{01} \\ \vdots \\ \zeta_{0,r+1} \end{bmatrix} \tag{3.9}$$

where ζ_{ij} is defined by

$$\begin{aligned} \zeta_{ij} &= Eu_{t_{m-i}} u_{t_{m-j}} \\ &= \gamma_{t_{m-i}-t_{m-j}} \end{aligned} \tag{3.10}$$

Now in the second stage it remains to show that w_n is also orthogonal to the remaining u_t for t=1,2,...,n-h-2. In order to prove this note that an alternative basis of the subspace S_{m-1} is the system which consists of u_{n-h-1}, $u_{n-h}, z_{t_{q+1}}, z_{t_{q+2}}, ..., z_{t_{m-1}}$. This follows straightforwardly from the definition of z_t. Hence, (3.8) can be written in an alternative manner as

$$w_n = u_n - \sum_{j=1}^{r-1} \mu_j z_{t_{m-j}} - \delta_{n,h} u_{n-h} - \delta_{n,h+1} u_{n-h-1} \tag{3.11}$$

where $\delta_{n,h}$ and $\delta_{n,h+1}$ can be obtained by multiplying both sides of (3.11) by u_{n-h-i} and taking expectations for i=0 and i=1, respectively. Since w_n is orthogonal to u_{n-h-i} we have $Ew_n u_{n-h-i}=0$ and since $Ez_{t_{m-j}} u_{n-h-i}=0$ (i=0,1 and j=1,2,...,r-1) as can be seen from (3.5) we obtain

$$\gamma_h = \delta_{n,h}\gamma_0 + \delta_{n,h+1}\gamma_1$$
$$\gamma_{h+1} = \delta_{n,h}\gamma_1 + \delta_{n,h+1}\gamma_0$$

Hence,

$$\begin{bmatrix}\delta_{n,h} \\ \delta_{n,h+1}\end{bmatrix} = \begin{bmatrix}\gamma_0 & \gamma_1 \\ \gamma_1 & \gamma_0\end{bmatrix}^{-1} \begin{bmatrix}\gamma_h \\ \gamma_{h+1}\end{bmatrix}$$
$$= \begin{bmatrix}\beta_{n,h} \\ \beta_{n,h+1}\end{bmatrix} \tag{3.12}$$

Now it follows from (3.12) and the definition of z_n that (3.11) can also be written as

$$w_n = u_n - \sum_{j=1}^{r-1} \mu_j z_{t_{m-j}} - \beta_{n,h} u_{n-h} - \beta_{n,h+1} u_{n-h-1}$$
$$= z_n - \sum_{j=1}^{r-1} \mu_j z_{t_{m-j}} \tag{3.13}$$

Hence, it follows from (3.5) and (3.13) that

$$Ew_n u_t = 0, \quad t=1,2,\ldots,n-h \tag{3.14}$$

Now it can be concluded from (3.7) and (3.14) that w_n is independent of the disturbance of any preceding available observation

$$Ew_n u_{t_j} = 0, \quad j=1,2,\ldots,m-1 \tag{3.15}$$

The last thing to be done is the evaluation of the variance of $w_n = w_{t_m}$. Define $\sigma_m^2 = Ew_{t_m}^2$ then it can be seen from (3.8) and (3.15) that

$$\sigma_m^2 = Eu_n w_n \tag{3.16}$$

Hence, after multiplying both sides of (3.8) by u_n and taking expectations we obtain in view of (3.16)

$$\sigma_m^2 = \gamma_0 - \sum_{j=1}^{r+1} \psi_j \gamma_{n-t_{m-j}} \tag{3.17}$$

Define $v_m = w_n/\sigma_m = (u_n - h_n)/\sigma_m$. Then v_m is orthogonal to all preceding disturbances u_t, i.e., $Ev_m u_{t_{m-j}} = 0$ $(j=1,\ldots,m-1)$ and $Ev_m^2 = 1$. Since v_m can be written as

$$v_m = (u_n - \psi_1 u_{t_{m-1}} - \ldots - \psi_{r+1} u_{t_{m-r-1}})/\sigma_m \tag{3.18}$$

the elements of the mth row of P are equal to

$$p_{mm} = \frac{1}{\sigma_m} \tag{3.19a}$$

$$p_{m,m-j} = -\frac{\psi_j}{\sigma_m}, \quad j=1,2,\ldots,r+1 \tag{3.19b}$$

$$p_{m,m-j} = 0, \quad j=r+2,r+3,\ldots,m-1 \tag{3.19c}$$

where the ψ_j and the σ_m can be determined from (3.9) and (3.17), respectively. The other m-1 rows can be determined in the same manner.

Finally, it should be noted that, in general, there will be no zeroes in the last row of P when, apart from possibly the first two observations, there is no other couple of two consecutive actual observations before observation n. Furthermore, the procedure can easily be generalized in a similar way for an arbitrary AR(p) process, $p \geq 3$.

If the disturbances u_t follow an AR(1) process with parameter φ then by a similar argument as that of the foregoing derivation it can be seen that (3.19) reduces to

$$p_{mm} = \frac{1}{\sigma_m} \tag{3.20a}$$

$$p_{m,m-1} = -\frac{\varphi_m}{\sigma_m}, \qquad \varphi_m = \varphi^{t_m - t_{m-1}} \tag{3.20b}$$

$$p_{m,m-j} = 0, \qquad j=2,\dots,m-1 \tag{3.20c}$$

where

$$\begin{aligned} \sigma_m^2 &= E(u_n - \varphi_m u_{n-n_m})^2, && n_m = t_m - t_{m-1} \\ &= \gamma_0 + \varphi_m^2 \gamma_0 - 2\varphi_m^2 \gamma_0 && (\text{note: } Eu_n u_{n-n_m} = \varphi_m \gamma_0,\ m>1) \\ &= (1-\varphi_m^2)/(1-\varphi^2), && \varphi_1 = 0 \end{aligned} \tag{3.21}$$

These results are identical with the results found by Wansbeek and Kapteyn (1985).

An interesting property of the covariance matrix of the disturbances of the m actual observations can be seen from (3.21). This property which is not mentioned in the article of Wansbeek and Kapteyn can be of interest when, for instance, the likelihood ratio test is applied to verify whether the disturbances are white noise.

Assume that n is an odd number and that the observations with an even rank number are missing then it follows from (3.21)

$$\sigma_m^2 = (1-\varphi^4)/(1-\varphi^2) = 1+\varphi^2 \tag{3.22}$$

Define $\Gamma_*=Eu_*u_*'$ then the determinant $|\Gamma_*|$ can be determined from (2.3.12) and (3.22)

$$|\Gamma_*| = \frac{(1+\varphi^2)^{m-1}}{1-\varphi^2} \qquad \text{(note: } \sigma_1^2=(1-\varphi^2)^{-1} \text{ and } m=(n+1)/2) \tag{3.23}$$

This seems to be somewhat peculiar because the determinant of the covariance matrix of the full set of AR(1) disturbances is equal to

$$|\Gamma| = |P|^{-2} = (1-\varphi^2)^{-1} \tag{3.24}$$

where P is the well-known Prais-Winsten transformation matrix. Expression (3.24) does not depend on n at all while $|\Gamma_*|$ does.

Also we have

$$\lim_{n\to\infty} |\Gamma_*|^{\frac{1}{n}} = (1+\varphi^2)^{\frac{1}{2}} \tag{3.25}$$

This implies that the Jacobian is no longer asymptotically negligible. So the Jacobian has to be fully taken into account when one applies, for instance, the likelihood ratio test for detecting autocorrelation except for the case where the number of missing observations is finite.

5.4 Estimation and test procedures

5.4.1 Estimation

Wansbeek and Kapteyn (1985) present a two step estimator for the AR(1) case which, in fact, is a Prais-Winsten procedure adjusted for missing observations. Their conclusion based on a Monte Carlo study, is that this two step estimator is almost as efficient as the maximum likelihood estimator and gives better results than the Cochrane-Orcutt procedure. These conclusions confirm the results found by Harvey and McAvinchey (1978) for the traditional model with AR(1) disturbances. They found from simulation experiments that the key feature of a successful estimation method is the use of the full GLS transform. Therefore, we propose a similar approach for the AR(2) case here. This means that the parameters φ_1 and φ_2 have to be determined from the OLS residuals from (2.4). This can be done in the following manner.

Define I as the index set of the rank numbers of the actual observations, i.e., $I=\{t_1,t_2,\ldots,t_m\}$ and define the subset I_k by

$$t_j \in I_k \quad \text{if} \quad t_j-k \in I, \quad k=0,1,2 \qquad (\text{note that } I_0=I) \tag{4.1}$$

Then estimate the autocovariances of u_t according to

$$\tilde{\gamma}_k = \frac{1}{n_k} \sum_{j\in I_k}^{n_k} \tilde{u}_{t_j}\tilde{u}_{t_j-k}, \quad k=0,1,2 \tag{4.2}$$

where n_k is the number of elements belonging to I_k and $\tilde{u}_{t_j}$ denotes the jth OLS residual from (2.4). Then estimate φ_1 and φ_2 according to the Yule-Walker equations

$$\begin{bmatrix} \tilde{\varphi}_1 \\ \tilde{\varphi}_2 \end{bmatrix} = \begin{bmatrix} \tilde{\gamma}_0 & \tilde{\gamma}_1 \\ \tilde{\gamma}_1 & \tilde{\gamma}_0 \end{bmatrix}^{-1} \begin{bmatrix} \tilde{\gamma}_1 \\ \tilde{\gamma}_2 \end{bmatrix} \tag{4.3}$$

Then the higher order autocovariances can be estimated according to (2.9)

$$\tilde{\gamma}_k = \tilde{\varphi}_1\tilde{\gamma}_{k-1} + \tilde{\varphi}_2\tilde{\gamma}_{k-2}, \quad k=3,4,\ldots \tag{4.4}$$

Now the lower triangular transformation matrix P can be calculated as described in the previous section and OLS estimation of the transformed model can be performed. This estimation method is asymptotically valid at least if the number of missing observations is finite. This means that the estimators of the regression coefficients of (2.5) are consistent and asymptotically efficient. And referring to the results of the experiments by Wansbeek and Kapteyn (op. cit.) it is expected that the performance of this estimation method will be reasonable in small samples.

In some cases, however, the approach described above is infeasible as Wansbeek and Kapteyn remark because of the pattern of the missing data. For instance, if all data or nearly all data with an even rank number are missing then $\tilde{\gamma}_1$ can not be calculated properly. In such cases it is advised to employ a maximum likelihood approach. Wansbeek and Kapteyn derive the first order conditions for the parameters of the first order autocorrelation case with missing observations according to the method presented in Magnus (1978). The estimates so obtained are asymptotically efficient provided the pattern of missing data is such that the parameters can be identified. However, it is cumbersome to derive these first order conditions for cases with higher order autocorrelation. Therefore, it is proposed here to derive straightforwardly the concentrated likelihood function in case of AR(2) disturbances.

For notational convenience we assume that there are M observations available rather than m. Then the log likelihood function of the actual observations can be written as

$$\ln(L) = -\frac{M}{2}\ln(2\pi) + \ln|P| - \frac{M}{2}\ln(\sigma_\varepsilon^2) - \frac{(Py_*-PX_*\alpha)'(Py_*-PX_*\alpha)}{2\sigma_\varepsilon^2} \tag{4.5}$$

This expression can be concentrated with respect to the parameter σ_ε^2 by taking the derivative with respect to σ_ε^2 and setting this equal to zero. This yields the ML estimator for σ_ε^2

$$\tilde{\sigma}_\varepsilon^2 = \frac{1}{M}(Py_*-PX_*\alpha)'(Py_*-PX_*\alpha) \tag{4.6}$$

Substituting (4.6) into (4.5) gives the concentrated log likelihood function

$$\ln(L_c) = -\frac{M}{2}\ln(2\pi) + \ln|P| - \frac{M}{2}\ln(\tilde{\sigma}_\varepsilon^2) - \frac{1}{2}M$$

$$= -\frac{M}{2}\{1+\ln(2\pi)\} - \frac{M}{2}\ln\left[\frac{\tilde{\sigma}_\varepsilon^2}{|P|^{2/M}}\right]$$

$$= -\frac{M}{2}\{1 + \ln(2\pi/M)\} - \frac{M}{2}\ln\left[\frac{(Py_*-PX_*\alpha)'(Py_*-PX_*\alpha)}{|P|^{2/M}}\right] \qquad (4.7)$$

Then upon substituting (3.19) into (4.7) we obtain the following expression

$$\ln(L_c) = -\frac{M}{2}\{1+\ln(2\pi/M)\} - \frac{M}{2}\ln(SSR(\varphi_1,\varphi_2)), \qquad (4.8a)$$

$$SSR(\varphi_1,\varphi_2) = \sum_{m=1}^{M}\left[\frac{\tilde{u}_{t_m} - \sum_{j=1}^{m-1}\psi_{mj}\tilde{u}_{t_{m-j}}}{\sigma_m/\bar{\sigma}}\right]^2 \qquad (4.8b)$$

$$\bar{\sigma} = |P|^{-1/M} = \left[\prod_{m=1}^{M}\sigma_m\right]^{1/M} \qquad (4.8c)$$

Note that ψ_{mj} is given by (3.9). To indicate the dependence of ψ_j on observation m a subscript m is added. In a full maximum likelihood procedure the elements in the M dimensional vector $\tilde{u}$ are to be taken according to

$$\tilde{u} = y_* - X_*(X_*'P'PX_*)^{-1}X_*'P'Py_* \qquad (4.9)$$

Since the ψ_{mj} are complicated functions of φ_1 and φ_2 it is hard to find expressions for the derivatives of ψ_{mj} with respect to φ_1 and φ_2. Provided that φ_1 and φ_2 can be identified, we can find by searching over a fine grid of feasible values of φ_1 and φ_2 the maximum likelihood estimates $\tilde{\varphi}_1$ and $\tilde{\varphi}_2$ for which $SSR(\varphi_1,\varphi_2)$ assumes its smallest value.

5.4.2 Tests for autocorrelation if observations are missing

5.4.2.1 The likelihood ratio test

In order to detect the appropriate type of second order autocorrelation one might carry out a sequence of the following three tests. The first test consists of

$$H_0: \varphi_1=\varphi_2=0 \quad \text{against} \quad H_1: \varphi_1\neq 0 \text{ and/or } \varphi_2\neq 0 \tag{4.10}$$

An asymptotically valid test among other tests is the likelihood ratio (LR) test. In view of (4.8a) the LR test statistic has the simple form

$$LR_1 = - M \ln \frac{SSR(\tilde{\varphi}_1,\tilde{\varphi}_2)}{SSR(0,0)} \tag{4.11}$$

where $\tilde{\varphi}_1$ and $\tilde{\varphi}_2$ can be obtained from (4.3) or from the maximum likelihood approach. LR_1 is asymptotically $\chi^2(2)$ under the null hypothesis $\varphi_1=\varphi_2=0$. If LR_1 takes an insignificant value one accepts H_0. If LR_1 is significant then one can perform a second test in a similar way

$$H_0: \varphi_2=0 \quad \text{against} \quad H_1: \varphi_2\neq 0 \tag{4.12}$$

This gives the following test statistic

$$LR_2 = - M \ln \frac{SSR(\tilde{\varphi}_1,\tilde{\varphi}_2)}{SSR(\tilde{\varphi}_1,0)} \tag{4.13}$$

which has a $\chi^2(1)$ distribution asymptotically. If this test statistic is insignificant one may conclude $\varphi_1\neq 0$ and $\varphi_2=0$. If LR_2 is significant one might test the following null hypothesis

$$H_0: \varphi_1=0 \text{ and } \varphi_2\neq 0 \quad \text{against} \quad H_1: \varphi_1\neq 0 \text{ and } \varphi_2\neq 0 \tag{4.14}$$

The corresponding test statistic is given by

$$LR_3 = -M \ln \frac{SSR(\tilde{\varphi}_1,\tilde{\varphi}_2)}{SSR(0,\tilde{\varphi}_2)} \tag{4.15}$$

An insignificant value of LR_3 indicates a pure AR(2) process while a significant value of LR_3 suggests $\varphi_1\neq 0$ and $\varphi_2\neq 0$.

5.4.2.2 The modified Lagrange multiplier (MLM) test

An alternative test which can be used, is the so called modified Lagrange multiplier (MLM) test adjusted for small samples. It is expected that in comparison with the LR test the MLM test gives better results in small samples, i.e., the actual significance level is relatively close to the chosen significance level; see Kiviet (1986). If all observations are available then the MLM test for an AR(2) process consists of an extension of the original regression equation with the lagged OLS residual vectors Be and B^2e, B being the backward shift operator (note: $B^k e_j=0, k\geqq j$). The MLM test, in fact, is a test on the joint significance of the estimated coefficients of Be and B^2e.

In order to see what the MLM test looks like in the case of missing observations the transformed model (2.5) is written in view of (3.19) in the following alternative way

$$\bar{y}_{t_m} = \bar{x}'_{t_m}\alpha + v_m, \qquad m=1,2,\ldots,M \tag{4.16}$$

$$v_m = \frac{u_{t_m} - \sum_{j=1}^{m-1} \psi_{mj} u_{t_{m-j}}}{\sigma_m} \tag{4.17}$$

Then it follows from (4.8) and (4.16) that the concentrated log likelihood function can be written as

$$\ln L(\Theta) = -\frac{M}{2}\ln\sum_{m=1}^{M}(\bar{\sigma}v_m)^2 - \frac{M}{2}\{1+\ln(2\pi/M)\} \tag{4.18}$$

where we have omitted the subscript c for notational convenience. If the parameter set is denoted by $\Theta=(\alpha',\varphi_1,\varphi_2)'$ then the ordinary Lagrange multiplier (LM) test statistic corresponding to the null hypothesis $\Theta=\Theta_0$ is equal to

$$LM = D\ln L(\Theta_0)'[-E\{D^2\ln L(\Theta_0)\}]^{-1}D\ln L(\Theta_0) \tag{4.19}$$

where D stands for the differential operator, i.e., $DL(\Theta)=\partial L(\Theta)/\partial\Theta$ and $D^2L(\Theta)=\partial^2L(\Theta)/\partial\Theta\partial\Theta'$. Asymptotically LM has a $\chi^2(s)$ distribution under H_0 where s is the number of elements in Θ; see Harvey (1986, p. 168). In order to avoid the evaluation of second derivatives in (4.19) we may use the two well-known equalities

$$E\{D\ln L(\Theta_0)\} = 0 \tag{4.20}$$

$$E\{D\ln L(\Theta_0)D\ln L(\Theta_0)'\} = -E\{D^2\ln L(\Theta_0)\} \tag{4.21}$$

see for a proof Section 7 of Chapter 6. Hence, the LM test statistic can be written as

$$LM = D\ln L(\Theta_0)'[\mathrm{cov}\{D\ln L(\Theta_0)\}]^{-1}D\ln L(\Theta_0) \tag{4.22}$$

From (4.18) it follows that $D\ln L(\Theta_0)$ is asymptotically equal to

$$\begin{aligned} D\ln L(\Theta_0) &= \frac{M}{2}(\Sigma\bar{\sigma}^2v_m^2)^{-1}2\Sigma\bar{\sigma}v_mz_m \\ &= (\Sigma v_mz_m)/(\bar{\sigma}s_0^2) \end{aligned} \tag{4.23}$$

where v_m, z_m and s_0^2 stand for

$$v_m = v_m(\Theta_0), \quad z_m = -\frac{\partial\{\bar{\sigma}(\Theta_0)v_m(\Theta_0)\}}{\partial\Theta} \quad \text{and} \quad s_0^2 = \frac{1}{M}\Sigma v_m^2(\Theta_0) \tag{4.24}$$

Assuming that v_m and z_m are uncorrelated it follows from (4.23) that the covariance matrix can be written asymptotically as

$$\begin{aligned} \text{cov}\{\text{Dln } L(\Theta_0)\} &= E(\Sigma v_m z_m)(\Sigma v_m z_m')/(\bar{\sigma}^2 s_0^4) \\ &= E(\Sigma v_m^2 z_m z_m')/(\bar{\sigma}^2 s_0^4) \\ &= \Sigma(z_m z_m')/(\bar{\sigma}^2 s_0^2) \end{aligned} \tag{4.25}$$

From (4.22), (4.23) and (4.25) it follows that for model (4.16) the LM test statistic becomes asymptotically

$$LM = s_0^{-2}\Sigma v_m z_m'(\Sigma z_m z_m')^{-1}\Sigma z_m v_m \tag{4.26}$$

This can be written in obvious matrix notation

$$\begin{aligned} LM &= v'Z(Z'Z)^{-1}Z'v/s_0^2 \\ &= \tilde{v}'\tilde{v}/s_0^2 \end{aligned} \tag{4.27}$$

where $\tilde{v}=\tilde{v}(\Theta_0)$ is defined as the vector of fitted values from a regression of $v_m(\Theta_0)$ on the derivatives of $-\bar{\sigma}v_m$ with respect to the elements in Θ evaluated in Θ_0 and denoted by $z_m(\Theta_0)$.

Remark 1. It should be borne in mind that the derivation of the LM test statistic as given by (4.27) is based on the assumptiom $Ev_m z_m=0$, i.e., the elements in the regressor vector z_m are predetermined. Therefore, when using (4.27) one has to verify whether $Ev_m z_m=0$.

Remark 2. From (4.8) it can easily be seen that maximizing the log likelihood function is equivalent to minimizing

$$\xi(\Theta) = \Sigma(\bar{\sigma}v_m)^2 \tag{4.28}$$

An alternative and more intuitive derivation of (4.27) can be based on (4.28). If $\Theta=\Theta_0$ is true then one might expect that the vector of the derivatives $D\xi_0=D\xi(\Theta_0)$ is approximately zero. In order to check this consider the test statistic

$$LM^* = (D\xi_0-ED\xi_0)'[\text{cov}(D\xi_0)]^{-1}(D\xi_0-ED\xi_0) \tag{4.29a}$$

where $D\xi_0$ can be derived from (4.28)

$$D\xi_0 = \partial\xi(\Theta_0)/\partial\Theta = -2\bar{\sigma}\Sigma v_m z_m = -2\bar{\sigma}Z'v \tag{4.29b}$$

Assuming once more $Ev_m z_m=0$ we have $ED\xi_0=0$ under the null $\Theta=\Theta_0$ whereas it follows from (4.29b) that $\text{cov}(D\xi_0)=E(D\xi_0)(D\xi_0)'$ is asymptotically equal to

$$\begin{aligned}\text{cov}(D\xi_0) &= 4\bar{\sigma}^2E\{Z'vv'Z\}\\ &= 4\bar{\sigma}^2s_0^2Z'Z\end{aligned} \tag{4.30}$$

Substituting (4.29b) and (4.30) into (4.29a) we obtain

$$LM^* = v'Z(Z'Z)^{-1}Z'v/s_0^2 \tag{4.31}$$

Assuming that the idempotent matrix $Z(Z'Z)^{-1}Z'$ has rank s we have $LM^*\sim\chi^2(s)$ under the null; see Theil (1971, p. 82). By comparing (4.31) with (4.27) it can easily be seen that $LM^*=LM$.

When only a small number of observations is available it is worthwhile to apply the MLM test. Besides, it is very easy to apply when the test only regards a subset of the parameters. Formula (4.27) suggests performing a regression of $v_m(\theta_0)$ on $z_m(\theta_0)$ and to test a hypothesis regarding a certain number of restrictions by using a standard F- or t-test; see Harvey (1986, pp. 173-174).

Suppose now that it is required to test the hypothesis of white noise disturbances in (2.1) against the AR(1) alternative. Neglecting the Jacobian $|P|=(\bar{\sigma})^{-M}$ we consider equation (4.16) in its unrestricted form. This becomes in view of (3.20) and (3.21)

$$\left[\frac{1-\varphi^2}{1-\varphi^{2n_m}}\right]^{\frac{1}{2}}\left[1-(\varphi B)^{n_m}\right]\left[Y_{t_m}-x'_{t_m}\alpha\right] = v_m, \quad n_m=t_m-t_{m-1} \text{ for } m>1 \tag{4.32}$$

or briefly

$$h(\varphi,m)u_{t_m} = v_m \tag{4.33}$$

The calculation of $v_m(\alpha,\varphi)$ under H_0 gives

$$\begin{aligned} v_m(\tilde{\alpha},0) &= h(0,m)\tilde{u}_{t_m} \\ &= \tilde{u}_{t_m} \end{aligned} \tag{4.34}$$

where $\tilde{u}_{t_m}$ is the OLS residual from (2.1). The derivatives of v_m are

$$\frac{\partial v_m(\tilde{\alpha},0)}{\partial\alpha} = -\,x_{t_m} \tag{4.35a}$$

$$\frac{\partial v_m(\tilde{\alpha},0)}{\partial\varphi} = \begin{cases} 0 & \text{if } n_m>1 \quad (4.35b) \\ -B(Y_{t_m}-\tilde{\alpha}X_{t_m}) = -\tilde{u}_{t_m-1} & \text{if } n_m=1 \quad (4.35c) \end{cases}$$

Hence, in the case of missing observations the MLM test against an AR(1) process consists of a regression of the OLS residuals $\tilde{u}_{t_m}$ from (2.1) on x'_{t_m} and s_{t_m-1}, where s_{t_m-1} is equal to $\tilde{u}_{t_m-1}$ if observation t_m-1 is available and zero if it is missing. The MLM test now becomes the Student's t-test of significance of the coefficient associated with s_{t_m-1}.

5.4.2.3 An infinite number of missing observations

In the case of daily data it is reasonable to assume that the number of missing data is proportional to the number of available data, because, for instance, the observations at Saturday and Sunday are missing. Consequently, the Jacobian of the transformation is no longer negligible; cf. the equations (3.22)-(3.25). In view of (3.20), (3.21) and (4.8) the log likelihood function becomes

$$\text{JLLF} = -\frac{M}{2}\ln\{\sum_{m=1}^{M}\bar{\sigma}^{2}v_m^{2}(\alpha,\varphi)\} \quad + \quad \text{const.} \tag{4.36}$$

where

$$\bar{\sigma}v_m(\alpha,\varphi) = \left[\prod_{j=1}^{M}\left[\frac{1-\varphi_j^2}{1-\varphi^2}\right]^{\frac{1}{2M}}\right]\left[\frac{1-\varphi^2}{1-\varphi_m^2}\right]^{1/2}(u_{t_m}-\varphi_m u_{t_{m-1}}) \tag{4.37a}$$

$$\varphi_m = \varphi^{n_m}, \qquad n_m = t_m - t_{m-1} \qquad \text{if } m>1 \tag{4.37b}$$

$$\varphi_1 = 0 \tag{4.37c}$$

In the sequel of this chapter we will write for notational convenience $\bar{\sigma}v_m(\alpha,\varphi)$ rather than the formally more correct expression $\bar{\sigma}(\alpha,\varphi)v_m(\alpha,\varphi)$; it is hoped it is not confusing. Since

$$\frac{\partial\bar{\sigma}v_m(\alpha,\varphi)}{\partial\varphi} = \frac{\partial\ln\{\bar{\sigma}v_m(\alpha,\varphi)\}}{\partial\varphi}\,\bar{\sigma}v_m(\alpha,\varphi) \tag{4.38}$$

and

$$\left. \frac{\partial \ln \frac{1-\varphi^n}{1-\varphi^2}}{\partial \varphi} \right|_{\varphi=0} = 0 \qquad \text{if } n \geq 2 \tag{4.39}$$

we get

$$\frac{\partial \bar{\sigma} v_m(\tilde{\alpha},0)}{\partial \varphi} = \bar{\sigma} v_m \left. \frac{-u_{t_m-1}}{u_{t_m} - \varphi u_{t_m-1}} \right|_{\substack{\alpha=\tilde{\alpha} \\ \varphi=0}} = - \tilde{u}_{t_m-1} \qquad \text{if } \quad n_m=1 \tag{4.40a}$$

$$= 0 \qquad \text{otherwise} \tag{4.40b}$$

Furthermore, we have

$$\frac{\partial \bar{\sigma} v_m(\tilde{\alpha},0)}{\partial \alpha} = - x_{t_m} \tag{4.41}$$

It follows from (4.40) and (4.41) that the MLM test of white noise disturbances against the AR(1) alternative is identical to the one in the case of a limited number of missing observations. In fact, this is due to the characteristic property of the Jacobian that its derivative is zero for $\varphi=0$. However, differences may arise when one tests the null hypothesis $\varphi=\varphi_0\neq 0$. Then the derivative of the Jacobian is no longer negligible.

Since the ψ_{mj} given by (3.19) are complicated functions of the parameters φ_1 and φ_2 it is tedious to derive directly a MLM test for the null hypothesis of AR(1) disturbances against the alternative of AR(2) disturbances. Therefore we will not pursue it here. It only should be noted that an alternative approach could be to calculate numerically the derivatives of v_m with respect to φ_1 and φ_2, respectively, and to evaluate these derivatives for the parameter estimates obtained by estimation under the null hypothesis.

5.4.2.4 The power of the MLM test

A peculiar phenomenon of the MLM test in the case of missing observations is as follows. Assume $t_m - t_{m-1} = 2$ for m=2,3,...,M and that there is an AR(1) process with parameter φ among the disturbances of the underlying model. Then it emerges that an MLM test of $\varphi=0$ based on the available data is impossible because

$$\frac{\partial v_m(\tilde{\alpha},0)}{\partial \varphi} = 0, \quad m=1,2,...,M \tag{4.42}$$

However, after renumbering the available observations as 1,2,...,M instead 1,3,...,2M-1 there is obviously an AR(1) process among the disturbances with parameter $\rho=\varphi^2$. An appropriate MLM test can easily be carried out. So there seems to be an MLM test paradox because the MLM test of $\varphi=0$ is inconsistent while the MLM test of $\rho=\varphi^2=0$ has power against the alternative $\rho\neq 0$.

The MLM test paradox might have consequences for the MLM test in the case that, for instance, observations 1,2,3,5,7...,2M-1 are available, giving M+1 observations. Rather than using all observations for the ordinary MLM test of $\varphi=0$ it seems to be better to ignore the second observation and to test $\rho=\varphi^2=0$. In cases where the pattern of missing observations is more complicated there is no evidence which test is to be preferred. It could be either the test of $\rho=0$ or the test of $\varphi=0$.

5.4.2.5 An adjusted Lagrange multiplier test

This subsection is concerned with the improvement of the power of the modified Lagrange multiplier test when data are missing. As we have seen in the previous subsection the MLM test may have a lack of power when many observations are missing according to a certain pattern. In this subsection it will be argued that the power can be improved by introducing an auxiliary parameter. To illustrate how this can be done consider the simple model with missing observations

$$y_{t_m} = x'_{t_m}\alpha + u_{t_m}, \qquad m=1,2,\dots,M \tag{4.43a}$$

$$u_t = \varphi u_{t-1} + \varepsilon_t \tag{4.43b}$$

Assume that apart from the first observation the index set I can be divided into two subsets I_1 and I_2, respectively,

$$t_m \in I_1 \quad \text{if} \quad t_m - t_{m-1} = 1 \tag{4.44a}$$

$$t_m \in I_2 \quad \text{if} \quad t_m - t_{m-1} = 2 \tag{4.44b}$$

and denote the number of elements in I_1 and I_2 by M_1 and M_2, respectively ($M_1+M_2=M$). Furthermore, define $\rho=\varphi^2$. Then in view of (4.36) and (4.37) the function to be minimized can be written as

$$\text{JSSR} = \sum_{t\in I_1} \bar{\sigma}^{-2} v_t^2(\delta) + \sum_{t\in I_2} \bar{\sigma}^{-2} v_t^2(\delta) \tag{4.45}$$

where

$$\bar{\sigma} v_t(\delta) = (1+\rho)^{\frac{M_2}{2M}}(u_t - \varphi u_{t-1}) \qquad \text{if } t \in I_1 \tag{4.46a}$$

$$\bar{\sigma} v_t(\delta) = (1+\rho)^{\left[\frac{M_2}{2M} - \frac{1}{2}\right]}(u_t - \rho u_{t-2})$$

$$= (1+\rho)^{-\frac{M_1}{2M}}(u_t - \rho u_{t-2}) \qquad \text{if } t \in I_2 \tag{4.46b}$$

with $\delta=(\alpha',\varphi,\rho)'$. Taking the derivatives of $\bar{\sigma} v_t(\delta)$ with respect to α, φ and ρ in $\delta_0=(\tilde{\alpha}',0,0)$ we obtain

$$\frac{\partial\bar{\sigma}v_t(\delta_0)}{\partial\alpha} = -x_t \tag{4.47a}$$

$$\frac{\partial\bar{\sigma}v_t(\delta_0)}{\partial\varphi} = -\tilde{u}_{t-1} \quad \text{if } t \in I_1 \tag{4.47b}$$

$$\frac{\partial\bar{\sigma}v_t(\delta_0)}{\partial\rho} = \frac{M_2}{2M}\tilde{u}_t \quad \text{if } t \in I_1 \tag{4.47c}$$

$$\frac{\partial\bar{\sigma}v_t(\delta_0)}{\partial\rho} = -\frac{M_1}{2M}\tilde{u}_t - \tilde{u}_{t-2} \quad \text{if } t \in I_2 \tag{4.47d}$$

Obviously $\partial\bar{\sigma}v_t(\delta_0)/\partial\rho$ depends upon $\tilde{u}_t=\bar{\sigma}v_t(\delta_0)$. Hence, a regression according to the MLM test is not appropriate; cf. the equations (4.28)-(4.31). In order to capture this problem consider the first order Taylor series expansion of $\bar{\sigma}v_t$ around δ_0. In view of (4.47) we obtain apart from asymptotically negligible terms

$$\bar{\sigma}v_t(\delta) = \tilde{u}_t - x_t'(\alpha-\tilde{\alpha}) - \varphi\tilde{u}_{t-1} + \rho\frac{M_2}{2M}\tilde{u}_t \qquad \text{if } t \in I_1 \tag{4.48a}$$

$$\bar{\sigma}v_t(\delta) = \tilde{u}_t - x_t'(\alpha-\tilde{\alpha}) - \rho\tilde{u}_{t-2} - \rho\frac{M_1}{2M}\tilde{u}_t \qquad \text{if } t \in I_2 \tag{4.48b}$$

Substitution of (4.48) into (4.45) yields the following function to be minimized

$$\text{JSSR} = \sum_{t\in I_1}\left[1+\rho\frac{M_2}{2M}\right]^2\left[\tilde{u}_t - \frac{x_t'(\alpha-\tilde{\alpha})+\varphi\tilde{u}_{t-1}}{1+(\rho M_2)/(2M)}\right]^2 +$$

$$+\sum_{t\in I_2}\left[1-\rho\frac{M_1}{2M}\right]^2\left[\tilde{u}_t - \frac{x_t'(\alpha-\tilde{\alpha})+\rho\tilde{u}_{t-2}}{1-(\rho M_1)/(2M)}\right]^2 \tag{4.49}$$

Expression (4.49) suggests ignoring the weights $[1+(\rho M_2)/(2M)]^2$ and $[1-(\rho M_1)/(2M)]^2$ which are equal to each other under H_0. Similar to the MLM test a test of $\varphi=0$ then consists of a joint significance test of $\varphi=\rho=0$ by means of the usual F-test in a regression of the OLS residuals $\tilde{u}_t$ from (4.43a) on x_t,

$s_1(t)$ and $s_2(t)$, where

$$s_1(t) = \tilde{u}_{t-1} \quad \text{if } t \in I_1 \tag{4.50a}$$

$$= 0 \quad \text{otherwise} \tag{4.50b}$$

$$s_2(t) = \tilde{u}_{t-2} \quad \text{if } t \in I_2 \tag{4.50c}$$

$$= 0 \quad \text{otherwise} \tag{4.50d}$$

The adjusted Lagrange multiplier test obviously employs the information contained in $\tilde{u}_{t-2}$ which might improve the power of the test compared to the modified Lagrange multiplier test.

5.5 Kalman filtering with missing observations

An alternative and interesting method for estimating regression models with missing observations and an arbitrary stationary ARMA(p,q) process among the disturbances is the Kalman filter method; see, for instance, Jones (1980) and Harvey and Pierse (1984). The state space representation of a regression model with ARMA (p,q) disturbances is given in Section 5 of Chapter 4; cf. (4.5.5)-(4.5.6). Since there is no measurement noise (R=0) the Kalman equations (4.3.32) can be written as

$$b_{t|t-1} = Ab_{t-1|t-1} \tag{5.1a}$$

$$P_{t|t-1} = AP_{t-1|t-1}A' + Q \tag{5.1b}$$

$$b_{t|t} = b_{t|t-1} + K_t(y_t - H_t b_{t|t-1}) \tag{5.1c}$$

$$P_{t|t} = (I - K_t H_t)P_{t|t-1} \tag{5.1d}$$

$$K_t = P_{t|t-1}H_t' f_t^{-1} \tag{5.1e}$$

$$f_t = H_t P_{t|t-1} H_t' \tag{5.1f}$$

The quantity f_t can be interpreted as the variance of the one step ahead prediction error. If y_t is missing and unknown this means that y_t has to be ignored in the recursive Kalman equations. This can simply be done by setting $f_t = \infty$ or, equivalently, $K_t=0$. This implies

$$P_{t|t} = P_{t|t-1} \tag{5.2a}$$

$$b_{t|t} = b_{t|t-1} \tag{5.2b}$$

The sequential Kalman filter algorithm can be applied once again for the following observations.

Applying the recursive Kalman equations also the log likelihood function of the observations $y_{t_1},\ldots,y_{t_m}$ can easily be evaluated given the parameter set $\{\varphi,\lambda\}$ of the covariance matrix of the disturbances. Hence, the likelihood ratio test can easily be carried out to test, for instance, the null hypothesis of AR(1) disturbances against the alternative of AR(2) disturbances. A minor problem of the use of the Kalman filter lies in finding the initial matrix $P_{0|0}$ which may be somewhat tedious; see Chapter 4. Another drawback of the Kalman filter equations is their recursive character which may lead to covariance matrices which are no longer nonnegative definite; see Harvey and Phillips (1979) and Kailath (1981, p. 172). The so called square root algorithm suggested by the authors does not suffer from this shortcoming. However, this algorithm might be rather time consuming.

Appendix 5.A Stationarity conditions for an AR(2) process

The stationarity conditions (2.3) follow straightforwardly from the corresponding first order Markov scheme representation

$$\begin{bmatrix} u_t \\ u_{t-1} \end{bmatrix} = \begin{bmatrix} \varphi_1 & \varphi_2 \\ 1 & 0 \end{bmatrix} \begin{bmatrix} u_{t-1} \\ u_{t-2} \end{bmatrix} + \begin{bmatrix} \varepsilon_t \\ 0 \end{bmatrix} \tag{A.1}$$

Stationarity requires that the eigenvalues of the transition matrix in (A.1) lie inside the unit circle. These eigenvalues are identical to the roots of the characteristic polynomial

$$x^2 - \varphi_1 x - \varphi_2 = 0 \tag{A.2}$$

If $\varphi_1^2+4\varphi_2 \geq 0$ then the stationarity condition implies

$$-2 < \varphi_1 \pm (\varphi_1^2+4\varphi_2)^{½} < 2 \tag{A.3}$$

or

$$(\varphi_1^2+4\varphi_2)^{½} < \varphi_1+2 \quad \text{and} \quad (\varphi_1^2+4\varphi_2)^{½} < 2-\varphi_1$$

$$\varphi_1^2+4\varphi_2 < (\varphi_1+2)^2, \quad \varphi_1^2+4\varphi_2 < (\varphi_1-2)^2 \quad \text{and} \quad |\varphi_1| < 2$$

$$4\varphi_2 < 4+4\varphi_1, \quad 4\varphi_2 < 4-4\varphi_1 \quad \text{and} \quad |\varphi_1| < 2 \tag{A.4}$$

It can easily be seen that (2.3a), (2.3b) and (2.3c) follow from (A.4) provided $\varphi_1^2+4\varphi_2 \geq 0$. In the case that $\varphi_1^2+4\varphi_2 < 0$ there are two conjugate complex roots. Hence, the stationarity condition becomes

$$|x_{12}|^2 = (\varphi_1^2 - \varphi_1^2 - 4\varphi_2)/4 = -\varphi_2 < 1 \tag{A.5}$$

Combining (A.4), (A.5), (2.3a) and (2.3b) yields (2.3c). Since all foregoing inequalities still hold if "<" is replaced by ">" apart from some minor modifications of the text, it can be concluded that the stationarity conditions given by (2.3) are necessary and sufficient.

CHAPTER VI

DISTRIBUTED LAG MODELS AND CORRELATED DISTURBANCES

6.1 Introduction

In this chapter we consider distributed lag models. In particular, we examine the case with one explanatory variable and geometrically declining lag coefficients. This special case is introduced in Section 2. A brief survey of the various estimation methods is given in Section 3. In Section 4 it is shown that Koyck's consistent two-step estimator can be determined uniquely from particular OLS results in a simple and direct way. Section 5 is concerned with an alternative estimation method which is asymptotically efficient. Also more general dynamic models with a stationary ARMA(p,q) error process are considered. Besides a small-sample adjustment of the estimation procedure is suggested. In Section 6 attention is paid to the estimation of distributed lag models with several distinct explanatory variables. Finally, Section 7 is concerned with the Cramér-Rao inequality and its relationship to the Pythagorean theorem.

6.2 The geometric distributed lag model

For simplicity we omit the constant term, and then the simple distributed lag model with geometrically declining coefficients can be written as

$$y_t = \beta x_t + \lambda\beta x_{t-1} + \lambda^2\beta x_{t-2} + \dots + u_t, \quad 0 \leq \lambda < 1 \tag{2.1}$$

or

$$y_t = \beta(1+\lambda B+\lambda^2 B^2+\dots)x_t + u_t = \frac{\beta}{1-\lambda B} x_t + u_t \qquad (2.2)$$

where B stands for the backward shift operator. Which estimation technique of the parameters β and λ is appropriate depends on the correlation structure of the disturbances. It is convenient to premultiply both sides of (2.2) by $1-\lambda B$, i.e., the so called Koyck transformation. This yields

$$y_t = \beta x_t + \lambda y_{t-1} + w_t, \qquad w_t = u_t - \lambda u_{t-1} \qquad (2.3)$$

In general, model (2.3) may arise in three distinct situations.

a. When it is assumed straightforwardly that the lag effects are declining in a geometric manner.

b. When the adaptive expectation assumption is made, that is,

$$x_t^e - x_{t-1}^e = \gamma(x_{t-1} - x_{t-1}^e), \qquad 0<\gamma<1 \qquad (2.4)$$

Among others Nerlove (1958) proposes (2.4) in the context of an agricultural supply function

$$y_t = \beta x_t^e + u_t \qquad (2.5)$$

where x_t^e stands for the expected future price x_t. Assumption (2.4) can also be written as

$$x_t^e = \frac{\gamma}{1-(1-\gamma)B} x_{t-1} \qquad (2.6)$$

and substitution of (2.6) into (2.5) yields

$$y_t = \frac{\beta\gamma}{1-(1-\gamma)B} x_{t-1} + u_t \tag{2.7}$$

which is of the same form as (2.2).

c. When the partial adjustment assumption is made. Here

$$y_t^* = \beta x_t + u_t \tag{2.8}$$

where y_t^* is the desired level of variable y given x_t and the adjustment of y_t will take place according to

$$y_t - y_{t-1} = \gamma(y_t^* - y_{t-1}), \quad 0<\gamma<1 \tag{2.9}$$

or

$$y_t^* = \frac{1-(1-\gamma)B}{\gamma} y_t \tag{2.10}$$

Combining (2.8) and (2.10) gives

$$y_t = \gamma\beta x_t + (1-\gamma)y_{t-1} + \gamma u_t \tag{2.11}$$

which is similar to (2.3) except for the disturbance term. Assuming u_t is white noise the OLS estimation of (2.11) is asymptotically efficient.

6.3 Estimation methods

Several estimation methods have been developed for the distributed lag model of the form (2.3). As a matter of fact OLS estimation produces consistent and asymptotically efficient results if the regressor x_t is fixed and the disturbances w_t follow a white noise process. However, problems arise when the disturbances follow, for instance, a simple first order Markov scheme

$$w_t = \varphi w_{t-1} + \varepsilon_t, \qquad -1<\varphi<1 \tag{3.1}$$

with ε_t a white noise process. Then the OLS-estimator is inconsistent, because the regressor y_{t-1} and the disturbance w_t are correlated; see Harvey (1986, p. 266). Several suggestions have been put forward in the literature to solve this problem. Liviatan (1963) proposes to use x_t and x_{t-1} as instrumental variables (IV). Though this IV estimator is consistent it is not efficient. Judge et al. (1980) suggest a two stage least squares approach which is more efficient. First they regress y_{t-1} on x_{t-1}, x_{t-2}, and so forth in order to obtain a calculated $\tilde{y}_{t-1}$. Then in a second step they regress y_t on x_t and $\tilde{y}_{t-1}$. Although this IV estimator is consistent it again is not fully efficient, since the method does not exploit the special structure of the correlations of the disturbances. When applying the Cochrane-Orcutt procedure it is convenient to write (2.3) as

$$y_t - \varphi y_{t-1} = \beta(x_t - \varphi x_{t-1}) + \lambda(y_{t-1} - \varphi y_{t-2}) + \varepsilon_t \tag{3.2}$$

which equation allows for a correlation structure of the form (3.1). Given an initial estimate of φ, the parameters β and λ in (3.2) can be estimated by OLS. Given these estimates the residuals $\tilde{w}_t$ from (2.3) can be calculated. Then a new estimate of φ can be obtained by regressing $\tilde{w}_t$ on $\tilde{w}_{t-1}$. This iterative procedure can be continued until the estimates achieve a specific degree of convergence. The estimates so obtained are consistent and asymptotically efficient; cf. Oberhofer and Kmenta (1974). However, in order to obtain the asymptotic covariance matrix of these estimators of β, λ and φ the full 3×3 information matrix has to be inverted since $E\{\partial \ln(L)/\partial\lambda\partial\varphi\}\neq 0$, L being the likelihood function; see Harvey (1986, p. 267).

If it is assumed that the disturbances u_t from the original model (2.1) are serially uncorrelated then the disturbances of model (2.3) follow an MA(1) process. Koyck (1954) considered this case. He suggests estimating (2.3) by OLS, yielding $\tilde{\beta}$ and $\tilde{\lambda}$, and to examine the asymptotic bias

$$\text{plim} \begin{bmatrix} \tilde{\beta} \\ \tilde{\lambda} \end{bmatrix} - \begin{bmatrix} \beta \\ \lambda \end{bmatrix} \tag{3.3}$$

In order to obtain a pair of consistent estimators $\tilde{\beta}_c$ and $\tilde{\lambda}_c$ one has to solve the following system of nonlinear equations

$$\begin{bmatrix} \tilde{\beta}_c \\ \tilde{\lambda}_c \end{bmatrix} = \begin{bmatrix} \Sigma x^2 & \Sigma xy_{-1} \\ \Sigma xy_{-1} & \Sigma y_{-1}^2 \end{bmatrix}^{-1} \begin{bmatrix} \Sigma xy \\ \Sigma y_{-1}y + \dfrac{\tilde{\lambda}_c \Sigma \tilde{w}_t^2}{1 + \tilde{\lambda}\tilde{\lambda}_c} \end{bmatrix} \tag{3.4}$$

Regarding this system of nonlinear equations Judge et al. (1980, p. 679) and Gupta (1971, p. 30) state that the computations to solve this nonlinear system are cumbersome. However, it will be shown in the following section, that the solution follows directly from the results of the OLS estimation of (2.3). Although Koyck's estimators $\tilde{\beta}_c$ and $\tilde{\lambda}_c$ are consistent for β and λ, respectively, they are not efficient; see Amemiya and Fuller (1967).

Klein (1958, p. 555) also considers the case where the disturbances from (2.1) are serially uncorrelated and he writes (2.1) as

$$y_t - u_t = \beta x_t + \lambda(y_{t-1} - u_{t-1}) \tag{3.5}$$

In this form we have a classical equation of the linear relation with variables subject to error. Since all the standard assumptions of the classical model are met in the case at hand Klein suggests minimizing

$$\xi = \Sigma u_t^2 + \Sigma u_{t-1}^2$$

$$= \Sigma(y_t - \beta x_t - \lambda\eta_{t-1})^2 + \Sigma(y_{t-1} - \eta_{t-1})^2, \qquad \eta_t = y_t - u_t \tag{3.6}$$

with respect to β,λ and each $\eta_t, t=2,3,\ldots,n$. The set of n+1 normal equations so obtained will after some algebraic manipulations lead to a quadratic equation in $\tilde{\lambda}_c$ which is identical with the quadratic implied by Koyck's system (3.4).

Gupta (1971, p. 30) pointed out that there is even a third and simpler manner to derive this quadratic. In short, he proves that minimizing

$$\frac{1}{1+\lambda^2}\Sigma(y_t - \beta x_t - \lambda y_{t-1})^2 \tag{3.7}$$

with respect to β and λ produces the same quadratic in $\tilde{\lambda}_c$ as given by Klein.

Amemiya and Fuller (1967) have derived the asymptotic covariance matrix of the estimators which are obtained by solving Koyck's quadratic and, consequently, Klein's and Gupta's quadratic. They have also shown that these estimators are less efficient as compared with the maximum likelihood estimator.

Under the assumption that the u_t have zero autocorrelation Klein (1958, p. 562) also shows how the maximum likelihood (ML) estimators for β and λ can be derived. He writes (2.1) as

$$y_t = \beta\sum_{j=0}^{t-1}\lambda^j x_{t-j} + \eta_0\lambda^t + u_t \tag{3.8}$$

$$\eta_0 = \beta\sum_{j=0}^{\infty}\lambda^j x_{-j} \quad (= y_0 - u_0) \tag{3.9}$$

The term $\eta_0\lambda^t$ is referred to as the truncation remainder. Given λ the parameters β and η_0 in (3.8) can be estimated consistently by OLS. Therefore, it is suggested to make a fine grid ($0<\lambda<1$) and to choose that value of λ

which minimizes the sum of squared residuals.

Dhrymes (1969) proposes an estimation technique which is also based on the ML principle. Besides, Dhrymes allows for the situation where the disturbances u_t in (2.1) or, equivalently, (3.8) are serially correlated

$$u_t = \varphi u_{t-1} + \varepsilon_t \tag{3.10}$$

Given λ and φ (3.8) can be written in matrix notation

$$y = X\gamma + u \tag{3.11}$$

where $\gamma=(\beta,\eta_0)'$ and X is a n×2 matrix. Dhrymes now premultiplies (3.11) by

$$P_a = |P|^{-1/n}P = (1-\varphi^2)^{-1/2n}P \tag{3.12}$$

where P is the well-known first order autocorrelation transformation matrix with parameter φ. Hence, we have

$$P_a y = P_a X\gamma + P_a u \tag{3.13}$$

or

$$y^* = X^*\gamma + u^* \tag{3.14}$$

The disturbances in (3.14) obey the usual assumptions. Hence, after simultaneous search in two dimensions over a permissible range of λ and φ the pair $(\tilde{\lambda},\tilde{\varphi})$ is chosen which produces the lowest sum of squared OLS residuals. It should be noted that since φ is unknown also the Jacobian $|P|$ of the transformation has to be taken into account for exact ML estimation of (2.1); see also Chapter 4, Section 2, viz. (4.2.16). Dhrymes's method gives consistent and asymptotically efficient estimators. The main drawback of this method lies in the fact that the two dimensional search procedure can be fairly time-consuming.

6.4 A simple formula for Koyck's consistent two-step estimator

In the context of the adaptive expectations hypothesis it is assumed that the disturbances of (2.3) follow an MA(1) process with parameter λ or, alternatively, that the disturbances in (2.1) are white noise. In practice this type of model is important; see Chow (1983, p. 103) and Harvey (1986, p. 226). Unfortunately, in such a case the OLS estimator of (2.3) is not consistent as we have seen in the previous section.

We now return to Koyck's system of nonlinear equations (3.4). Koyck proves that solving this system we obtain a consistent estimator of the parameters of the adaptive expectations model. In this section it is shown how the solution of this system of nonlinear equations can easily be obtained from the OLS estimation of (2.3). In obvious matrix notation (2.3) can be written as

$$y = X\delta + w \tag{4.1}$$

with $\delta=(\beta,\lambda)'$. Then the OLS estimators of (2.3) are

$$\begin{aligned} \begin{bmatrix} \tilde{\beta} \\ \tilde{\lambda} \end{bmatrix} &= (X'X)^{-1}X'y \\ &= \begin{bmatrix} \beta \\ \lambda \end{bmatrix} + VX'w \end{aligned} \tag{4.2}$$

where V denotes $(X'X)^{-1}$. Furthermore, we have in view of (2.1) and (2.3)

$$\text{plim}\ (\tfrac{1}{n}\ X'w) = \text{plim}\ \frac{1}{n}\begin{bmatrix}\Sigma x_t w_t\\ \Sigma y_{t-1}w_t\end{bmatrix}$$

$$= \begin{bmatrix}0\\ Eu_{t-1}(u_t-\lambda u_{t-1})\end{bmatrix} = \begin{bmatrix}0\\ -\lambda\end{bmatrix}\sigma_u^2 \qquad (4.3)$$

where $\sigma_u^2=Eu_t^2$. Combining (4.2) and (4.3) yields

$$\text{plim}\begin{bmatrix}\tilde{\beta}\\ \tilde{\lambda}\end{bmatrix} = \begin{bmatrix}\beta\\ \lambda\end{bmatrix} + \text{plim}\ nV\begin{bmatrix}0\\ -\lambda\end{bmatrix}\sigma_u^2$$

$$= \begin{bmatrix}\beta\\ \lambda\end{bmatrix} - \lambda\ \text{plim}\begin{bmatrix}nv_{12}\sigma_u^2\\ nv_{22}\sigma_u^2\end{bmatrix} \qquad (4.4)$$

In order to eliminate the unknown σ_u^2 consider the sum of the squared OLS residuals from (2.3)

$$\tilde{w}'\tilde{w} = w'(I - X(X'X)^{-1}X')w$$

$$= w'w - w'XVX'w \qquad (4.5)$$

From (4.2) and (4.5) it can be seen that

$$\tilde{w}'\tilde{w} = w'w - w'X\begin{bmatrix}\tilde{\beta}-\beta\\ \tilde{\lambda}-\lambda\end{bmatrix} \qquad (4.6)$$

After dividing both sides of (4.6) by n and taking plims of both sides we obtain

$$\text{plim}\ (s_w^2) = \sigma_w^2 + \text{plim}\ (0,\ \lambda\sigma_u^2)\begin{bmatrix}\tilde{\beta}-\beta\\ \tilde{\lambda}-\lambda\end{bmatrix}$$

$$= \sigma_w^2 + \lambda\sigma_u^2\ \text{plim}\ (\tilde{\lambda}-\lambda) \qquad (4.7)$$

where use has been made of (4.3), and where s_w^2 denotes the OLS estimator of σ_w^2, the variance of w_t. Since

$$\sigma_w^2 = (1 + \lambda^2)\, \sigma_u^2 \tag{4.8}$$

we obtain upon substituting (4.8) into (4.7)

$$\sigma_u^2 = \text{plim} \frac{s_w^2}{1 + \tilde{\lambda}\lambda} \tag{4.9}$$

Now substitution of (4.9) into (4.4) yields

$$\text{plim} \begin{bmatrix} \tilde{\beta} \\ \tilde{\lambda} \end{bmatrix} = \begin{bmatrix} \beta \\ \lambda \end{bmatrix} - \lambda\, \text{plim} \frac{n}{1 + \tilde{\lambda}\lambda} \begin{bmatrix} v_{12} s_w^2 \\ v_{22} s_w^2 \end{bmatrix} \tag{4.10}$$

Noting that $v_{22} s_w^2$ is equal to the estimated variance of the OLS estimator of $\tilde{\lambda}$ (s_λ^2) and, similarly, $v_{12} s_w^2$ is equal to the estimated covariance of $\tilde{\lambda}$ and $\tilde{\beta}$ ($s_{\lambda\beta}$) we can write (4.10) as

$$\text{plim}\, \tilde{\beta} = \beta - \lambda\, \text{plim} \frac{n}{1 + \tilde{\lambda}\lambda}\, s_{\lambda\beta} \tag{4.11a}$$

$$\text{plim}\, \tilde{\lambda} = \lambda - \lambda\, \text{plim} \frac{n}{1 + \tilde{\lambda}\lambda}\, s_\lambda^2 \tag{4.11b}$$

Now it emerges from (4.11b) that a consistent estimator $\tilde{\lambda}_c$ for λ equals the positive root of the following quadratic equation

$$\tilde{\lambda}_c(1 + \tilde{\lambda}\tilde{\lambda}_c) - ns_\lambda^2\tilde{\lambda}_c - \tilde{\lambda}(1 + \tilde{\lambda}\tilde{\lambda}_c) = 0 \tag{4.12a}$$

$$\tilde{\lambda}\tilde{\lambda}_c^2 + (1 - ns_\lambda^2 - \tilde{\lambda}^2)\tilde{\lambda}_c - \tilde{\lambda} = 0 \tag{4.12b}$$

$$\tilde{\lambda}_c^2 + \mu\tilde{\lambda}_c - 1 = 0, \qquad \mu = (1 - ns_\lambda^2 - \tilde{\lambda}^2)/\tilde{\lambda} \tag{4.12c}$$

From the last equation it is obvious that the product of the two roots equals -1. Since $0<\lambda<1$ the largest root of (4.12) is a consistent estimator for λ. So we have

$$\tilde{\lambda}_c = -\frac{1}{2}\mu + \frac{1}{2}(\mu^2 + 4)^{½} \tag{4.13}$$

From (4.11a) it can be seen that a consistent estimator of β equals

$$\tilde{\beta}_c = \tilde{\beta} + n\tilde{\lambda}_c \frac{s_{\lambda\beta}}{(1 + \tilde{\lambda}\tilde{\lambda}_c)} \tag{4.14}$$

After evaluating $\tilde{\lambda}_c$ and $\tilde{\beta}_c$ one might calculate consistent residuals from (2.3). A further analysis of these residuals might be carried out in order to check whether the hypothesis of white noise disturbances in (2.1) is plausible or not.

6.5 Efficient estimation of dynamic models

6.5.1 Introduction

In this section we consider models where the lagged dependent variable is one of the explanatory variables whereas the disturbances follow a stationary ARMA(p,q) process. For these models the GLS estimator based on an estimated autocovariance matrix of the disturbances is no longer efficient; see Theil (1971, p. 415). The estimation approach in this section resembles the estimation procedures which are already proposed in Durbin (1960), Amemiya and Fuller (1967), Hatanaka (1974) and Harvey (1986). One of the common points in the contributions of these authors is that they use in the second stage of the estimation procedure the lagged residuals as an explanatory variable, particularly if the disturbances follow an AR(1) process. In the presentation of their results the authors differ in the way they show that their estimators are efficient.

In Harvey (1986, pp. 269-271) it is explained how one step of the iterative Gauss-Newton optimization procedure can be used to obtain efficient estimators if consistent estimators are already available. Rather than the Gauss-Newton procedure we prefer in this section a simple first order Taylor series expansion to arrive at the same results; see also Harvey (1986, pp. 133-134). Because of the convention in the literature we will use the name Gauss-Newton estimator although only use will be made of a first order Taylor series expansion.

There are three major differences between the approach in this section and the approaches of the authors mentioned above. Firstly, in order to obtain consistent estimators we propose in this section as well as in Section 6 to exploit Klein's ML estimation method described in Section 3. Secondly, the derivation given here, of the Gauss-Newton two (three) step method being efficient is slightly different from those given by the authors mentioned above. Thirdly, we suggest some small sample adjustments in order to improve the efficiency of the estimators in the case of a small sample.

6.5.2 An efficient 3-step Gauss-Newton estimation method

Consider the dynamic model

$$y_t = \beta x_t + \lambda y_{t-1} + w_t \tag{5.1}$$

where w_t is a stationary ARMA(p,q) process. In this subsection we will describe a consistent and efficient 3-step Gauss-Newton estimator.

Step 1. As we have seen in the previous section (5.1) can be written as

$$y_t = \beta \sum_{j=0}^{t-1} \lambda^j x_{t-j} + \eta_0 \lambda^t + u_t, \quad u_t = \frac{w_t}{1-\lambda B} \tag{5.2}$$

$$\eta_0 = \sum_{j=0}^{\infty} \lambda^j x_{-j} \tag{5.3}$$

Given an initial value λ_0 (5.2) can be approximated by using the first order

Taylor series expansion for λ^j. Since we have

$$\lambda^j \approx \lambda_0^j + (\lambda-\lambda_0)j\lambda_0^{j-1} = (1-j)\lambda_0^j + \lambda j\lambda_0^{j-1} \tag{5.4}$$

we can write (5.2) as

$$y_t = \beta \sum_{j=0}^{t-1} (1-j)\lambda_0^j x_{t-j} + \beta\lambda \sum_{j=1}^{t-1} j\lambda_0^{j-1} x_{t-j} + \eta_0\lambda_0^t(1-t) + \eta_0\lambda t\lambda_0^{t-1} + u_t \tag{5.5}$$

For simplicity the remainder is neglected. Equation (5.5) is a nonlinear regression model with restrictions on the parameters. However, β and η_0 given, λ can be estimated by OLS. Moreover, after replacing λ_0 by the new estimate of λ the parameter λ can be estimated once again by OLS, and so forth until convergence to a certain degree is attained. Next, given λ, the parameters β and η_0 in (5.2) can be estimated by OLS. Given these new estimates of β and η_0, λ can be estimated in a second round in the same manner as described above. Due to the results found by Oberhofer and Kmenta (1974) it can easily be seen that the estimation results from such an iterative zig-zag procedure are identical with the estimates which are found by minimizing

$$\Sigma u_t^2 = \Sigma u_t^2(\lambda,\beta,\eta_0) \tag{5.6}$$

Finally, it should be noted that the second explanatory variable on the right hand side of (5.2) is asymptotically negligible; see Schmidt (1975).

<u>Step 2</u>. After inserting the consistent but inefficient least squares estimates for β and λ in (5.1) the residuals $\tilde{w}_t$ can be calculated. By means of the calculated residuals $\tilde{w}_t$ an appropriate ARMA(p,q) process for the disturbances in (5.1) has to be identified and its parameters have to be estimated in a consistent manner by one of the usual methods.

Step 3. After consistent estimates of the error process have been obtained an asymptotically efficient estimator can be derived by exploiting the Taylor series expansion once more. Assume, for instance, that the w_t follow an ARMA(1,1) process

$$w_t = \varphi w_{t-1} + \varepsilon_t - \Theta\varepsilon_{t-1} \tag{5.7}$$

Then (5.1) can be written as

$$y_t = \beta x_t + \lambda y_{t-1} + \varphi(y_{t-1}-\beta x_{t-1}-\lambda y_{t-2}) - \Theta\varepsilon_{t-1} + \varepsilon_t \tag{5.8}$$

By employing (5.7) the white noise process ε_t can be approximated by

$$e_t = \tilde{w}_t - \tilde{\varphi}\tilde{w}_{t-1} + \tilde{\Theta}e_{t-1}, \qquad t\geqq 3 \text{ (set } e_2=0) \tag{5.9}$$

where $\tilde{\varphi}$ and $\tilde{\Theta}$ are the consistent estimators of φ and Θ, respectively. Denote the vector of parameters to be estimated in (5.8) by α and write (5.8) briefly as $y_t=f_t(\alpha)+\varepsilon_t$. Then the Taylor series expansion of the right hand side of (5.8) around the consistent estimator $\tilde{\alpha}=(\tilde{\beta},\tilde{\lambda},\tilde{\varphi},\tilde{\Theta})'$ becomes

$$y_t = \tilde{y}_t + \tilde{\varphi}\tilde{w}_{t-1} - \tilde{\Theta}e_{t-1} + (\beta-\tilde{\beta})\frac{\partial f_t(\tilde{\alpha})}{\partial\beta} + (\lambda-\tilde{\lambda})\frac{\partial f_t(\tilde{\alpha})}{\partial\lambda} + (\varphi-\tilde{\varphi})\frac{\partial f_t(\tilde{\alpha})}{\partial\varphi} +$$

$$+ (\Theta-\tilde{\Theta})\frac{\partial f_t(\tilde{\alpha})}{\partial\Theta} + \varepsilon_t, \qquad t\geqq 3 \tag{5.10a}$$

or briefly

$$e_t = \frac{\partial f_t(\tilde{\alpha})}{\partial\alpha'}(\alpha-\tilde{\alpha}) + \varepsilon_t, \quad t\geqq 3 \tag{5.10b}$$

The derivatives of f_t with respect to the four parameters can be calculated in a recursive way as can be seen from (5.8)

$$\frac{\partial f_t(\tilde{\alpha})}{\partial\beta} = x_t - \tilde{\varphi}x_{t-1} + \tilde{\Theta}\,\frac{\partial f_{t-1}(\tilde{\alpha})}{\partial\beta}, \quad t\geqq 2 \quad (\text{note: } \frac{\partial f_t}{\partial\alpha} = -\frac{\partial\varepsilon_t}{\partial\alpha}) \tag{5.11a}$$

$$\frac{\partial f_t(\tilde{\alpha})}{\partial\lambda} = y_{t-1} - \tilde{\varphi}y_{t-2} + \tilde{\Theta}\,\frac{\partial f_{t-1}(\tilde{\alpha})}{\partial\lambda}, \quad t\geqq 3 \tag{5.11b}$$

$$\frac{\partial f_t(\tilde{\alpha})}{\partial\varphi} = \tilde{w}_{t-1} + \tilde{\Theta}\,\frac{\partial f_{t-1}(\tilde{\alpha})}{\partial\varphi}, \quad t\geqq 3 \tag{5.11c}$$

$$\frac{\partial f_t(\tilde{\alpha})}{\partial\Theta} = -\,e_{t-1} + \tilde{\Theta}\,\frac{\partial f_{t-1}(\tilde{\alpha})}{\partial\Theta}, \quad t\geqq 3 \tag{5.11d}$$

For t=1 or t=2 all four initial derivatives are set equal to zero.

In the specific case $\Theta=0$ (5.10) reduces to

$$y_t-\tilde{\varphi}y_{t-1} = \beta(x_t-\tilde{\varphi}x_{t-1}) + \lambda(y_{t-1}-\tilde{\varphi}y_{t-2}) + (\varphi-\tilde{\varphi})\tilde{w}_{t-1} + \varepsilon_t, \quad t\geqq 3 \tag{5.12}$$

This equation is also given in Hatanaka (1974). He considers model (5.1) with an AR(1) process among the disturbances. For this specific case he proves that the OLS estimation of (5.12) yields a consistent and efficient estimator.

Next, we will show in a slightly different way that the OLS estimation of (5.10) yields consistent estimators in more general cases as well. Denote the vector of regressors in (5.10b) by $\tilde{z}_t(=\partial f_t(\tilde{\alpha})/\partial\alpha=-\partial\varepsilon_t(\tilde{\alpha})/\partial\alpha)$ and the corresponding matrix of regressors by $\tilde{Z}=Z(\tilde{\alpha})$. Then it follows from (5.10b) that

$$\tilde{\alpha}_{OLS} - \tilde{\alpha} = \alpha - \tilde{\alpha} + (\tilde{Z}'\tilde{Z})^{-1}\tilde{Z}'\varepsilon \tag{5.13}$$

Since all the regressors in (5.10) are predetermined, as can be seen from (5.11), we have asymptotically

$$\sqrt{n}(\tilde{\alpha}_{OLS}-\alpha) \sim N(0,\, \sigma^2 \mathrm{plim}(\tilde{Z}'\tilde{Z}/n)^{-1}) \tag{5.14}$$

In order to prove that $\tilde{\alpha}_{OLS}$ is an efficient estimator we consider more precisely the information matrix $-E\{D^2 \ln L(\alpha)\}$ for the case at hand, D being the differential operator, i.e., $DL(\alpha)=\partial L(\alpha)/\partial\alpha$ and $D^2L(\alpha)=\partial^2 L(\alpha)/\partial\alpha\partial\alpha'$. Since the Jacobian can be ignored asymptotically the log likelihood function is asymptotically equal to

$$\ln L(\alpha) = -\frac{1}{2}n \ln(\sigma^2) - \frac{1}{2\sigma^2}\Sigma\varepsilon_t^2 + \text{const.} \tag{5.15}$$

cf. Harvey (1986, pp. 103-106). Hence, we have

$$D\ln L(\alpha) = -\frac{1}{\sigma^2}\Sigma\frac{\partial\varepsilon_t}{\partial\alpha}\varepsilon_t = \frac{1}{\sigma^2}\Sigma\frac{\partial f_t}{\partial\alpha}\varepsilon_t = \frac{1}{\sigma^2}Z'\varepsilon \tag{5.16}$$

Now making use of the well-known equality

$$E\{D\ln L(\alpha)D\ln L(\alpha)'\} = -E\{D^2\ln L(\alpha)\} \tag{5.17}$$

it follows from (5.16) that

$$E(Z'Z)/\sigma^2 = -E\{D^2\ln L(\alpha)\} \tag{5.18}$$

and as $\tilde{\alpha}$ is a consistent estimator for α it follows from (5.14) and (5.18) that the asymptotic covariance matrix of $\sqrt{n}(\tilde{\alpha}_{OLS}-\alpha)$ equals

$$\text{plim}\left(\frac{\tilde{Z}'\tilde{Z}}{n}\right)^{-1}\sigma^2 = \lim_{n\to\infty} -n\left[E\{D^2\ln L(\alpha)\}\right]^{-1} \tag{5.19}$$

Although we considered σ^2 in (5.15) to be fixed it should be noted that σ^2 is also an unknown parameter to be estimated. However, σ^2 does not play a role in the α-block of the inverse of the information matrix since the common part in the Hessian has zero expectation as can be seen from (5.11) and (5.16)

$$E \frac{\partial^2}{\partial\sigma^2\partial\alpha} \ln L(\alpha,\sigma^2) = E \frac{\partial}{\partial\sigma^2} \text{DlnL}(\alpha) = -\frac{1}{\sigma^2} E\{\text{DlnL}(\alpha)\} = 0 \qquad (5.20)$$

Therefore, it follows from (5.14), (5.19) and the Cramér-Rao inequality that $\tilde{\alpha}_{OLS}$ is an efficient estimator; see for a proof of (5.17) and the Cramér-Rao inequality Section 7 of this chapter.

Remark. The relationship of (5.13) to the iterative Gauss-Newton procedure is immediately apparent. A step of the Gauss-Newton procedure leads to a similar expression. For the minimization of $\Sigma\varepsilon_t^2(\alpha)$ we have according to Gauss-Newton

$$\alpha_{k+1} = \alpha_k + (Z'Z)^{-1}Z'\varepsilon(\alpha_k), \quad Z = Z(\alpha_k) = \frac{-\partial\varepsilon(\alpha_k)}{\partial\alpha'} \qquad (5.21)$$

where $\varepsilon(\alpha_k)$ stands for the vector of residuals corresponding to $\alpha=\alpha_k$. In other words, $\alpha_{k+1}-\alpha_k$ can be seen as the OLS estimate in the regression of $\varepsilon(\alpha_k)$ on the columns of $Z(\alpha_k)$; note that this corresponds with the Taylor series expansion $\varepsilon(\alpha_{k+1})\approx\varepsilon(\alpha_k)+(\partial\varepsilon(\alpha_k)/\partial\alpha')(\alpha_{k+1}-\alpha_k)$. Harvey (1986, p. 134 and p. 270) gives a different explanation for the relationship to the two step Gauss-Newton estimator. The procedure described here can also be generalized for the case that the disturbances of (5.1) follow an ARMA(p,q) process.

6.5.3 A Gauss-Newton-Prais-Winsten estimation method with small sample adjustments

It can be seen from (5.10) and (5.11) that inaccuracies can be introduced in the OLS estimates of (5.10) for two reasons. Firstly, the second observation is neglected. This can entail a substantial loss in efficiency; see Harvey and McAvinchey (1978). Secondly, the initial derivatives in (5.11) can wrongly be set equal to zero for t=1 and t=2. This might affect the approximation of the Taylor series expansion on the right hand side of (5.10) for low values of t, especially for t=3.

In order to take into account the information contained in the first two observations in a more efficient manner it seems to be worthwhile to consider the following regression model with n-1 observations

$$p_{11}y_2 = \beta p_{11}x_2 + \lambda p_{11}y_1 + p_{11}w_2 \tag{5.22a}$$

$$p_{21}y_2 + p_{22}y_3 = \beta(p_{21}x_2+p_{22}x_3) + \lambda(p_{21}y_1+p_{22}y_2) + (p_{21}w_2+p_{22}w_3) \tag{5.22b}$$

$$e_t + \tilde{z}_t'\tilde{\alpha} = \tilde{z}_t'\alpha + \varepsilon_t, \quad t\geq 4 \tag{5.22c}$$

The last equation is equivalent to (5.10b). The lower triangular 2×2 matrix P can be seen as the Cholesky decomposition of the inverse of the covariance matrix of w_1 and w_2, i.e., $\Gamma_2^{-1} = P'P$; note $E\varepsilon_t^2=1$. The elements of P depend on φ and Θ and can be determined from (2.3.9), (2.4.11) and (2.4.12). This yields

$$p_{11} = \gamma_0^{-1/2} \tag{5.23a}$$

$$p_{21} = -\gamma_1/(\gamma_0\sigma_2) \tag{5.23b}$$

$$p_{22} = \sigma_2^{-1} \tag{5.23c}$$

$$\sigma_2 = (\gamma_0 - \gamma_1^2\,\gamma_0^{-1})^{1/2} \tag{5.23d}$$

where γ_0 and γ_1 follow from (2.7.8)

$$\gamma_0 = (1+\Theta^2-2\varphi\Theta)/(1-\varphi^2) \tag{5.24a}$$

$$\gamma_1 = (1-\varphi\Theta)(\varphi-\Theta)/(1-\varphi^2) \tag{5.24b}$$

Since the true parameters are unknown, the matrix P has to be estimated by replacing φ and Θ in (5.24) by consistent estimators $\tilde{\varphi}$ and $\tilde{\Theta}$. Setting the regressors corresponding to φ and Θ equal to zero for t=2,3 the parameters β,λ,φ and Θ can be estimated by OLS estimation of (5.22).

A further refinement can be realized by defining

$$\begin{bmatrix}\zeta_2\\ \zeta_3\end{bmatrix} = P \begin{bmatrix}y_2-\beta x_2-\lambda y_1\\ y_3-\beta x_3-\lambda y_2\end{bmatrix} = P \begin{bmatrix}w_2\\ w_3\end{bmatrix} \tag{5.25}$$

Since P depends upon φ and θ, ζ_2 and ζ_3 are functions of φ and θ as well. After calculating $\partial\zeta_t(\tilde{\alpha})/\partial\varphi$ and $\partial\zeta_t(\tilde{\alpha})/\partial\theta$ for t=2,3 and, consequently, the corresponding part of the regressor vectors $\tilde{z}_2$ and $\tilde{z}_3$, that is

$$\tilde{z}_2' = (\tilde{p}_{11}x_2,\ \tilde{p}_{11}y_1,\ -\partial\zeta_2(\tilde{\alpha})/\partial\varphi,\ -\partial\zeta_2(\tilde{\alpha})/\partial\theta)' \tag{5.26a}$$

$$\tilde{z}_3' = (\tilde{p}_{21}x_2+\tilde{p}_{22}x_3,\ \tilde{p}_{21}y_1+\tilde{p}_{22}y_2,\ -\partial\zeta_3(\tilde{\alpha})/\partial\varphi,\ -\partial\zeta_3(\tilde{\alpha})/\partial\theta)' \tag{5.26b}$$

the regression equations to be estimated by OLS can be written as

$$\zeta_2(\tilde{\alpha}) + \tilde{z}_2'\tilde{\alpha} = \tilde{z}_2'\alpha + \zeta_2 \tag{5.27a}$$

$$\zeta_3(\tilde{\alpha}) + \tilde{z}_3'\tilde{\alpha} = \tilde{z}_3'\alpha + \zeta_3 \tag{5.27b}$$

$$e_t + \tilde{z}_t'\tilde{\alpha} = \tilde{z}_t'\alpha + \varepsilon_t, \quad 4\leqq t\leqq n \tag{5.27c}$$

To illustrate what (5.27a and b) look like in the case of MA(1) disturbances we derive the first observational equation for the MA(1) case. Equation (5.25) becomes

$$\zeta_2 = (1+\theta^2)^{-1/2}(y_2-\beta x_2-\lambda y_1) \qquad (\text{note: } p_{11} = \gamma_0^{-1/2}=(1+\theta^2)^{-1/2}) \tag{5.28}$$

A Taylor expansion of the right hand side of (5.28) gives

$$\zeta_2 = (1+\tilde{\theta}^2)^{-1/2}(y_2-\tilde{\beta}x_2-\tilde{\lambda}y_1) - (\beta-\tilde{\beta})(1+\tilde{\theta}^2)^{-1/2}x_2 - (\lambda-\tilde{\lambda})(1+\tilde{\theta}^2)^{-1/2}y_1 + \\ - (\theta-\tilde{\theta})\tilde{\theta}(1+\tilde{\theta}^2)^{-1\frac{1}{2}}\tilde{w}_2 \tag{5.29}$$

Upon rearranging (5.29) we obtain

$$(1+\tilde{\theta}^2)^{-\frac{1}{2}} y_2 = \beta(1+\tilde{\theta}^2)^{-\frac{1}{2}} x_2 + \lambda(1+\tilde{\theta})^{-\frac{1}{2}} y_1 + (\theta-\tilde{\theta})\tilde{\theta}(1+\tilde{\theta}^2)^{-1\frac{1}{2}} \tilde{w}_2 + \zeta_2 \tag{5.30}$$

A somewhat peculiar property of equation (5.30) is that the regressor $\tilde{w}_2$ is obviously correlated with the disturbance $\zeta_2=(1+\theta^2)^{-\frac{1}{2}}w_2$. This is due to the use of the ML estimation method. In order to shed some light on this point consider the Gauss-Newton optimization procedure for the simple model with AR(1) disturbances

$$y_t = \beta x_t + w_t, \qquad t=1,2,\ldots,n \tag{5.31}$$

$$w_t = \varphi w_{t-1} + \varepsilon_t \tag{5.32}$$

As we have seen in Chapter 4, cf. (4.2.16), maximizing the concentrated log likelihood function of $y_1,\ldots,y_n$ is equivalent to minimizing

$$\sum_{t=1}^{n} \zeta_t^2, \qquad \zeta_1 = |P|^{-1/n}(1-\varphi^2)^{\frac{1}{2}} w_1$$

$$\zeta_t = |P|^{-1/n}(y_t - \varphi y_{t-1} - \beta x_t + \beta\varphi x_{t-1}), \quad t\geqq 2$$

$$|P| = (1-\varphi^2)^{\frac{1}{2}} \tag{5.33}$$

with respect to α, α being defined by $\alpha=(\beta,\varphi)'$. According to the iterative Gauss-Newton procedure one has to regress $\zeta_t(\tilde{\alpha})$ on its derivatives. Discarding the first observation the derivatives are equal to

$$\frac{\partial\zeta_t(\tilde{\alpha})}{\partial\beta} = -(1-\tilde{\varphi}^2)^{-1/2n}(x_t-\tilde{\varphi}x_{t-1}), \quad t\geqq 2 \tag{5.34a}$$

$$\frac{\partial\zeta_t(\tilde{\alpha})}{\partial\varphi} = -(1-\tilde{\varphi}^2)^{-1/2n}\tilde{w}_{t-1} + \frac{\tilde{\varphi}}{n}(1-\tilde{\varphi}^2)^{-\frac{1}{2n}-1}(\tilde{w}_t-\tilde{\varphi}\tilde{w}_{t-1}), \quad t\geqq 2 \tag{5.34b}$$

After some lines of straightforward algebra it can be shown that the Gauss-Newton procedure is equivalent to the repeated OLS estimation of a regression equation of the type

$$y_t - \tilde{\varphi} y_{t-1} = \beta(x_t - \tilde{\varphi} x_{t-1}) + (\varphi - \tilde{\varphi})(\tilde{w}_{t-1} - \frac{\tilde{\varphi}}{n(1-\tilde{\varphi}^2)} e_t) + (1-\tilde{\varphi}^2)^{1/2n} \zeta_t, \quad (5.35a)$$

$$e_t = \tilde{w}_t - \tilde{\varphi}\tilde{w}_{t-1} \quad (5.35b)$$

The latter equation suggests that the Gauss-Newton procedure might imply that one of the regressors and the disturbance term have nonzero covariance. However, the estimators obtained by the iterative Gauss-Newton procedure are ML estimators provided that the procedure converges.

A final point concerns the calculation of the derivatives $\partial f_t(\tilde{\alpha})/\partial\alpha$ by means of (5.11). As we have seen the recursive calculations in (5.11) are started up by setting the initial derivatives equal to zero. Probably the efficiency of the OLS estimator of (5.27) can be improved by taking a kind of steady state solutions of (5.11a and b) for t=2. This yields

$$\partial f_2(\tilde{\alpha})/\partial\beta = (x_2 - \tilde{\varphi} x_1)/(1-\tilde{\theta}) \quad (5.36a)$$

$$\partial f_2(\tilde{\alpha})/\partial\lambda = (y_2 - \tilde{\varphi} y_1)/(1-\tilde{\theta}) \quad (5.36b)$$

The question whether (5.36a), for instance, is a good approximation of (5.11a) depends on the following two issues:

i) Is $\sum_{j=1}^{\infty} \theta^j (x_{2-j} - \varphi x_{1-j})$ different from zero?

ii) Is $\sum_{j=1}^{\infty} \theta^j \{(x_2 - \varphi x_1) - (x_{2-j} - \varphi x_{1-j})\}$ approximately zero?

If both answers are in the affirmative then (5.36) should be used. This might be the case especially if one of the predetermined variables is subject to a strong trend.

6.6 Dynamic models with several geometric distributed lags

In this section we will propose an estimation method that can be applied to models with two lagged dependent variables. The method resembles Klein's approach for deriving the ML estimators for model (2.1) by means of the so called truncation remainder. Here we will introduce two truncation remainders in order to obtain consistent estimates. These consistent estimates can be used in a 3-step Gauss-Newton estimation procedure to obtain efficient estimators in the same manner as described in the previous section. Also models with more complicated lag structures are discussed.

Consider a model with two explanatory variables with different geometric distributed lag structures, i.e.,

$$y_t = \beta \frac{x_t}{1-\psi_1 B} + \gamma \frac{z_t}{1-\psi_2 B} + u_t \tag{6.1}$$

where u_t follows a stationary ARMA(p,q) process. Multiplying both sides of (6.1) by $(1-\psi_1 B)(1-\psi_2 B)$ yields

$$y_t = \beta x_t - \beta\psi_2 x_{t-1} + \gamma z_t - \gamma\psi_1 z_{t-1} + \lambda_1 y_{t-1} + \lambda_2 y_{t-2} + w_t \tag{6.2}$$

$$\lambda_1 = \psi_1 + \psi_2 \tag{6.3}$$

$$\lambda_2 = -\psi_1\psi_2 \tag{6.4}$$

$$w_t = (1-\lambda_1 B-\lambda_2 B^2)u_t \tag{6.5}$$

For the sake of simplicity we consider first the model

$$y_t = \beta x_t + \lambda_1 y_{t-1} + \lambda_2 y_{t-2} + w_t \tag{6.6}$$

The stationarity conditions are

$$\lambda_2 + \lambda_1 < 1 \tag{6.7a}$$

$$\lambda_2 - \lambda_1 < 1 \tag{6.7b}$$

$$\lambda_2 > -1 \tag{6.7c}$$

These conditions follow straightforwardly from the condition that the roots of

$$1 - \lambda_1 B - \lambda_2 B^2 = 0 \tag{6.8}$$

lie outside the unit circle; see Section 2 of Chapter 5. Furthermore, we drop the condition that the roots of (6.8) are real valued. As before we denote the roots of (6.8) by ψ_1^{-1} and ψ_2^{-1}. Then (6.6) can be written as

$$\begin{aligned} y_t &= \frac{\beta x_t}{1-\lambda_1 B-\lambda_2 B^2} + u_t, \qquad u_t = \frac{w_t}{1-\lambda_1 B-\lambda_2 B^2} \\ &= \frac{\beta x_t}{(1-\psi_1 B)(1-\psi_2 B)} + u_t \\ &= \beta\Big(\sum_{j=0}^{\infty} \psi_1^j B^j\Big)\Big(\sum_{j=0}^{\infty} \psi_2^j B^j\Big) x_t + u_t \end{aligned} \tag{6.9}$$

Because of the absolute convergence of the geometric series we can take the Cauchy product on the right hand side of (6.9). This yields

$$y_t = \beta \sum_{j=0}^{\infty} \Big(\sum_{i=0}^{j} \psi_1^i \psi_2^{j-i} \Big) B^j x_t + u_t$$

$$= \beta \sum_{j=0}^{\infty} \frac{\psi_1^{j+1} - \psi_2^{j+1}}{\psi_1 - \psi_2} x_{t-j} + u_t \quad \text{if } \psi_1 \neq \psi_2 \tag{6.10}$$

$$= \beta \sum_{j=0}^{\infty} (j+1)\psi_1^j x_{t-j} + u_t \quad \text{if } \psi_1 = \psi_2 \tag{6.11}$$

In the sequel we neglect the case that $\psi_1=\psi_2$. Now similar to Koyck's model (see (3.8)) two truncation remainders can be introduced, viz.

$$y_t = \beta \sum_{j=0}^{t-1} \frac{\psi_1^{j+1} - \psi_2^{j+1}}{\psi_1 - \psi_2} x_{t-j} + \mu_1\psi_1^t - \mu_2\psi_2^t + u_t \tag{6.12}$$

$$\mu_1 = \beta \sum_{j=0}^{\infty} \frac{\psi_1^{j+1}}{\psi_1 - \psi_2} x_{-j} \tag{6.13a}$$

$$\mu_2 = \beta \sum_{j=0}^{\infty} \frac{\psi_2^{j+1}}{\psi_1 - \psi_2} x_{-j} \tag{6.13b}$$

If ψ_1 and ψ_2 are complex valued then $\psi_2 = \bar{\psi}_1$ and they can be written as

$$\psi_1 = \rho\ (\cos\varphi + i\ \sin\varphi) \tag{6.14a}$$

$$\psi_2 = \rho\ (\cos\varphi - i\ \sin\varphi) \tag{6.14b}$$

Then by De Moivre's theorem we get

$$\frac{\psi_1^{j+1} - \psi_2^{j+1}}{\psi_1 - \psi_2} = \rho^j \frac{\sin(j+1)\varphi}{\sin\varphi} \tag{6.15}$$

Hence, in the case of complex valued roots (6.12) becomes

$$y_t = \beta \sum_{j=0}^{t-1} \rho^j \frac{\sin(j+1)\varphi}{\sin\varphi} x_{t-j} + \eta_0(t)\rho^t + u_t \tag{6.16}$$

where

$$\eta_0(t) = \beta \sum_{j=0}^{\infty} \rho^j \frac{\sin(j+1+t)\varphi}{\sin\varphi} x_{-j} \tag{6.17}$$

Since $\sin(a+b)=\sin(a)\cos(b)+\sin(b)\cos(a)$ (6.17) can be written as

$$\eta_0(t) = \eta_1\cos(t\varphi) + \eta_2\sin(t\varphi) \tag{6.18}$$

where

$$\eta_1 = \beta \sum_{j=0}^{\infty} \rho^j \frac{\sin(j+1)\varphi}{\sin\varphi} x_{-j} \tag{6.19a}$$

$$\eta_2 = \beta \sum_{j=0}^{\infty} \rho^j \frac{\cos(j+1)\varphi}{\sin\varphi} x_{-j} \tag{6.19b}$$

Substitution of (6.18) into (6.16) yields

$$y_t = \beta \sum_{j=0}^{t-1} \rho^j \frac{\sin(j+1)\varphi}{\sin\varphi} x_{t-j} + \eta_1\rho^t\cos(t\varphi) + \eta_2\rho^t\sin(t\varphi) + u_t \tag{6.20}$$

As described in the previous section consistent estimates for β, ψ_1, ψ_2 can be obtained by least squares estimation of (6.12) or (6.20) by employing the Taylor series expansion around estimates of ψ_1 and ψ_2 in an iterative manner. In the case that ψ_1 and ψ_2 are real valued one should estimate (6.12) whereas (6.20) is to be estimated if ψ_1 and ψ_2 are complex valued. Initial estimates of ψ_1 and ψ_2 can be determined from the estimates of λ_1 and λ_2 which can be obtained by IV estimation of (6.6) using x_{t-1} and x_{t-2} as instrumental variables. As a matter of fact ρ and φ have to be taken if the estimates of ψ_1 and ψ_2 are complex valued. Alternatively, one might search over the permissible range of ψ_1 and ψ_2 and choose those values for which the sum of squared residuals of (6.12) or (6.20) assumes the smallest value.

The second step consists of the calculation of the residuals from

$$\tilde{w}_t = y_t - \tilde{\lambda}_1 y_{t-1} - \tilde{\lambda}_2 y_{t-2} - \tilde{\beta} x_t \tag{6.21}$$

Then identify the ARMA process of the $\tilde{w}_t$ and estimate the parameters consistently. In the third step consistent and asymptotically efficient estimates of the parameters in (6.6) can be obtained along the same lines as described in the previous section.

Finally, for the sake of completeness we will consider the more general dynamic model with two explanatory variables and a more general lag structure

$$y_t = \beta \frac{x_t}{\psi_1(B)} + \gamma \frac{z_t}{\psi_2(B)} + u_t \tag{6.22}$$

where u_t is a stationary ARMA(p,q) process. If ψ_1 and ψ_2 are polynomials in B with degrees p_1 and p_2, respectively, it is hard to follow the same approach as described above. In particular, it is tedious to apply Klein's approach in order to obtain consistent estimates. Therefore, we propose an alternative approach. If we assume for simplicity that $p_1+p_2=3$ then (6.22) can be transformed according to Koyck

$$y_t = \lambda_1 y_{t-1} + \lambda_2 y_{t-2} + \lambda_3 y_{t-3} + \beta\psi_2(B)x_t + \gamma\psi_1(B)z_t + w_t \tag{6.23a}$$

$$w_t = \lambda(B)u_t = \psi_1(B)\psi_2(B)u_t \tag{6.23b}$$

$$\lambda(B) = 1 - \lambda_1 B - \lambda_2 B^2 - \lambda_3 B^3 \tag{6.23c}$$

In order to obtain consistent estimates the following instrumental variables (IV) approach can be followed. The approach strongly resembles the approach advocated by Judge et al. (p. 666). Regress y_t on x_t, z_t, x_{t-1}, z_{t-1}, and so forth and include as many lags in x_t and z_t as appear to make significant increases in the overall correlation for the multiple regression. Then calculate $\tilde{y}_t$, and use $\tilde{y}_{t-1}$, $\tilde{y}_{t-2}$ and $\tilde{y}_{t-3}$ as instrumental variables for the IV estimation of (6.23a). After consistent estimates for $\lambda_1,\lambda_2,\lambda_3,\beta,\gamma$ and the coefficients of the polynomials ψ_1 and ψ_2 have been obtained, the residuals $\tilde{w}_t$ can be calculated. Next, the same efficient estimation technique can be applied as described in Section 5, especially the steps 2 and 3.

It should be noted that it is not necessary to include λ_1, λ_2 and λ_3 in the regression. We can write (6.23a) as

$$\psi_1(B)\psi_2(B)y_t = \beta\psi_2(B)x_t + \gamma\psi_1(B)z_t + w_t \tag{6.24}$$

Assuming that w_t can be written as $\Theta(B)\varepsilon_t/\mu(B)$ we can write (6.24) as

$$y_t = \{1-\psi_1(B)\psi_2(B)\mu(B)\}y_t + \beta\psi_2(B)\mu(B)x_t + \gamma\psi_1(B)\mu(B)z_t + \Theta(B)\varepsilon_t \quad (6.25a)$$

$$= f_t(\alpha) + \varepsilon_t \quad (6.25b)$$

Next the derivatives $\partial f_t(\tilde{\alpha})/\partial\alpha$ required for the first order Taylor series expansion can be calculated in a similar manner as (5.11).

6.7 The Cramér-Rao inequality and the Pythagorean theorem

In this section we will give a derivation of the Cramér-Rao inequality with special emphasis on its geometric interpretation and its relationship to regression theory, especially to the well-known inequality $R^2 \leq 1$. The derivation itself resembles the one given by Cramer (1986, p. 19). However, the underlying geometric and Pythagorean interpretations are quite different. Although in the derivation given here use wil be made of geometric concepts already described in preceding chapters the text is rather self-contained. The main point to be borne in mind is that in a linear space of stochastic variables the inner product of two stochastic elements x and y is defined as $(x,y)=Exy$ whereas the norm or length is defined as $\|x\|=(x,x)^{1/2}=(Ex^2)^{1/2}$. In order to shed some light on the definition of the inner product of stochastic variables note that the correlation coefficient of two stochastic variables with zero expectation is defined as

$$\rho(x,y) = \frac{Exy}{\{Ex^2\ Ey^2\}^{1/2}} = \frac{(x,y)}{\{(x,x)\cdot(y,y)\}^{1/2}} = \frac{(x,y)}{\|x\|\cdot\|y\|} = \cos(x,y) \quad (7.1)$$

whereas the correlation coefficient of a sample of pairs of observations $(x_1,y_1),\ldots,(x_n,y_n)$ can be written as

$$\rho_{xy} = \frac{\Sigma x_i y_i}{\{\Sigma x_i^2 \ \Sigma y_i^2\}^{1/2}} = \frac{(x,y)}{\{(x,x)\cdot(y,y)\}^{1/2}} = \frac{(x,y)}{\|x\|\cdot\|y\|} = \cos(x,y) \tag{7.2}$$

where $x=(x_1,\ldots,x_n)'$ and $y=(y_1,\ldots,y_n)'$ and where x_i and y_i are measured as deviations from their respective means. The similarity between (7.1) and (7.2) illustrates that the operators E and Σ can be employed for a similar role in both statistics and linear algebra. Furthermore, when two stochastic variables x and y have zero inner product $(x,y)=Exy=0$ then we call them orthogonal to each other; they also have zero covariance provided at least one of them has zero mean.

In the sequel of this section we will denote the likelihood function $L(\alpha,y_1,\ldots,y_n)$ of the observations $y_1,\ldots,y_n$ by $L(\alpha,y)$, $\partial./\partial\alpha$ by D, $\partial./\partial\alpha'$ by D', $\partial./\partial\alpha\partial\alpha'$ by D^2, $\int_{-\infty}^{\infty}\ldots\int_{-\infty}^{\infty}$ by $\int$, and $dy_1\ldots dy_n$ by dy. Finally, we denote a matrix with elements $\int a_{ij}dy$ by $\int Ady$.

Assuming that the regularity conditions are satisfied such that D and $\int$ can be interchanged we have

$$E\{D\ln L(\alpha,y)\} = \int \frac{DL(\alpha,y)}{L(\alpha,y)}\, L(\alpha,y)dy = D\int L(\alpha,y)dy = D1 = 0 \tag{7.3}$$

If $\tilde{\alpha}$, more precisely $\tilde{\alpha}(y)$, denotes an unbiased estimator of the column vector α of parameters we can write $\tilde{\alpha}-\alpha$ as a sum of two parts: a part linear in the elements of $D\ln L(\alpha,y)$ and a part which consists of random variables with zero expectation stacked in a residual vector

$$\tilde{\alpha}-\alpha = K\, D\ln L(\alpha,y) + u \tag{7.4}$$

or

$$y_1 = K\, y_2 + u \tag{7.5}$$

It is recalled from Section 2.2 of Chapter 4 that $K_i D\ln L(\alpha,y)$ can be interpreted as the orthogonal projection of $\tilde{\alpha}_1-\alpha_1$ onto the m-dimensional space generated by the elements in $D\ln L(\alpha,y)$ where K_i stands for the ith row of the so called Kalman matrix K of regression coefficients and m is the number of elements in α. As we have seen the scalar $E\|u\|^2=E\{\Sigma u_i^2\}=tr\{Euu'\}=Eu'u$ is minimized by solving the following m^2 normal equations $Euy_2'=E(y_1-Ky_2)y_2'=0$, yielding the so called least squares solution

$$K = Ey_1y_2'[Ey_2y_2']^{-1} \tag{7.6}$$

Making use of (7.3) we find for the covariance matrix of $y_1=\tilde{\alpha}-\alpha$ and $y_2=D\ln L(\alpha,y)$

$$\begin{aligned} E\{(\tilde{\alpha}-\alpha)D\ln L(\alpha,y)'\} &= E\{\tilde{\alpha}\ D\ln L(\alpha,y)'\} = \int \tilde{\alpha}\ \frac{DL(\alpha,y)'}{L(\alpha,y)} L(\alpha,y)dy \\ &= D'\int \tilde{\alpha}\ L(\alpha,y)\ dy = D'\alpha = I_m \quad (\text{note: } \tilde{\alpha}=\tilde{\alpha}(y)) \end{aligned} \tag{7.7}$$

Hence, by (7.6) and (7.7)

$$K = [Ey_2y_2']^{-1} = [E\{D\ln L(\alpha,y)D\ln L(\alpha,y)'\}]^{-1} \tag{7.8}$$

It should be emphasized that by this choice of K all elements of the so called orthogonal remainder u are orthogonal to all elements in $D\ln L(\alpha,y)$ hence, by (7.3), (7.4) and (7.8) we have

$$\begin{aligned} cov(\tilde{\alpha}-\alpha) &= cov\{K\ D\ln L(\alpha,y)\} + cov(u) \\ &= K\ K^{-1}K' + cov(u) \\ &= K + cov(u) \end{aligned} \tag{7.9}$$

Since cov(u) is positive semidefinite the covariance matrix of $\tilde{\alpha}-\alpha$ always exceeds $K=[cov\{D\ln L(\alpha,y)\}]^{-1}$ by a positive semidefinite matrix. This is known as the Cramér-Rao inequality. According to the Pythagorean theorem we can also write

$$\|\tilde{\alpha}_i-\alpha_i\|^2 = \|K_i D\ln L(\alpha,y)\|^2 + \|u_i\|^2, \quad i=1,\dots,m \tag{7.10}$$

where $\|x\|^2$ stands for Ex^2 or in an alternative manner as

$$E\|\tilde{\alpha}-\alpha\|^2 = E\|KD\ln L(\alpha,y)\|^2 + E\|u\|^2 \tag{7.11}$$

where $\|x\|^2$ stands for Σx_i^2. It is even possible to write

$$\|\tilde{\alpha}-\alpha\|^2 = \|KD\ln L(\alpha,y)\|^2 + \|u\|^2 \tag{7.12}$$

provided that the inner product of two column vectors x and y with stochastic elements is defined by $(x,y)=E\{\Sigma x_i y_i\}=\text{tr}\{Exy'\}$ while the corresponding norm in this specific Euclidean space is defined by $\|x\|=(x,x)^{½}=[E\{\Sigma x_i^2\}]^{½}=\{\text{tr}(Exx')\}^{½}$.

In fact, the Cramér-Rao inequality can be seen as a specific application of the well-known inequality $R^2\leq 1$ in multiple regression analysis where R is the multiple correlation coefficient; see Theil (1971, pp. 164-165, especially his equations (1.3) to (1.5)) for a geometric interpretation of R^2 and its relationship to the Pythagorean theorem. For the case at hand we could define the m×m diagonal matrix R^2 of squares of multiple correlation coefficients with elements

$$R_{ii}^2 = \frac{K_{ii}}{K_{ii}+Eu_i^2} = \frac{K_{ii}}{E(\tilde{\alpha}_i-\alpha_i)^2} = \frac{K_{ii}}{\|\tilde{\alpha}_i-\alpha_i\|^2} = 1 - \frac{\|u_i\|^2}{\|\tilde{\alpha}_i-\alpha_i\|^2} \tag{7.13}$$

Then the Cramér-Rao inequality can be written in obvious matrix notation as

$$0 \leq R^2 \leq I_m \qquad \text{or} \qquad 0 \leq tr(R^2) \leq m \tag{7.14}$$

Remark 1. For the sake of completeness it should be noted that many authors set K equal to the inverse of the information matrix $K=I^{-1}=[-E\{D^2 lnL(\alpha,y)\}]^{-1}$ rather than $K=[E\{DlnL(\alpha,y)DlnL(\alpha,y)'\}]^{-1}$. This is based on the well-known equality

$$\begin{aligned} E\{D^2 lnL(\alpha,y)\} &= \int D'\left[\frac{DL(\alpha,y)}{L(\alpha,y)}\right]L(\alpha,y)dy \\ &= \int \frac{D^2L(\alpha,y)}{L(\alpha,y)} L(\alpha,y)dy - \int \frac{DL(\alpha,y)DL(\alpha,y)'}{L(\alpha,y)^2} L(\alpha,y)dy \\ &= D^2 1 - E\{DlnL(\alpha,y)DlnL(\alpha,y)'\} \\ &= - E\{DlnL(\alpha,y)DlnL(\alpha,y)'\} \end{aligned} \tag{7.15}$$

Remark 2. At first glance the choice of the regressors in the regression equation (7.4) may look somewhat arbitrary. But the rationale behind (7.4) is a simple first order Taylor series expansion of $DlnL(\tilde{\alpha},y)$ around α, yielding

$$DlnL(\tilde{\alpha},y) \approx DlnL(\alpha,y) + D^2 lnL(\alpha,y)(\tilde{\alpha}-\alpha) \tag{7.16}$$

Taking into account the first order condition $DlnL(\tilde{\alpha},y)=0$ we obtain

$$\tilde{\alpha}-\alpha \approx [-D^2 lnL(\alpha,y)]^{-1} DlnL(\alpha,y) \tag{7.17}$$

or in view of (7.15)

$$\tilde{\alpha}-\alpha \approx [E\{DlnL(\alpha,y)DlnL(\alpha,y)'\}]^{-1} DlnL(\alpha,y) \tag{7.18}$$

These two equations suggest that the elements in $\tilde{\alpha}-\alpha$ can be approximated by a linear combination of the elements in $D\ln L(\alpha,y)$ and this is reflected by regression equation (7.4).

Remark 3. Note that (7.17) and (7.18), respectively, are the rationale behind the two iterative methods for the maximization of the log likelihood function $\ln L(\alpha,y)$

$$\alpha_{k+1} = \alpha_k - [D^2\ln L(\alpha_k,y)]^{-1} D\ln L(\alpha_k,y) \tag{7.19}$$

$$\alpha_{k+1} = \alpha_k + \left[\sum_{t=1}^{n} D\ln L(\alpha_k,y_t) D\ln L(\alpha_k,y_t)'\right]^{-1} \sum_{t=1}^{n} D\ln L(\alpha_k,y_t) \tag{7.20}$$

The recursions (7.19) and (7.20) are referred to as the Newton-Raphson method and the method of Berndt, Hall, Hall, and Hausman, respectively; note that $D\ln L(\alpha,y_t)D\ln L(\alpha,y_t)'$ is always singular for $m>1$ whereas, in general, its expectation or summation is not. Furthermore, we have $D\ln L(\alpha,y)=\Sigma D\ln L(\alpha,y_t)$ if the observations $y_1,\ldots,y_n$ are independent. See also Harvey (1986, p. 128 and p. 135).

CHAPTER VII

TEST STRATEGIES FOR DISCRIMINATING BETWEEN AUTOCORRELATION AND MISSPECIFICATION

7.1 Introduction

The main aim of the Durbin-Watson (DW) test is to detect an AR(1) process among the disturbances in a linear regression model. However, it is well known that a significant value of the DW test statistic, dw, can be caused by various types of misspecification, viz., AR(1) disturbances, non-AR(1) autocorrelated disturbances (NAR) and misspecification of the regression function (MS), i.e., explanatory variables are wrongly omitted or the functional form of the model is incorrect. Since a wrong diagnosis of the problem may lead to a wrong correction procedure and, consequently, to poor estimation results it is of great interest to have available a test strategy which is capable of discriminating between the three types of misspecification mentioned above.

Although several authors (Sargan,1964; Savin and White, 1978b) have paid attention to tests for discriminating between AR(1) and NAR or between AR(1) and MS, Thursby (1981) claims to be the first author who presents a general test strategy for discriminating between the case that there is no problem (NP) with the regression and the three alternatives mentioned above. His test strategy is partly based on the work of the earlier authors. However, Knottnerus (1985) pointed out that Thursby's test strategy can be criticized for several distinct reasons. Referring to Porter and Kashyap (1984) also Godfrey (1987) criticizes Thursby's test strategy and he proposes an alternative test strategy that does not suffer from the problems described by Porter and Kashyap, and Knottnerus.

The plan of this chapter is as follows. In Section 2 Thursby's test strategy is described. In Section 3 Knottnerus's critical notes on Thursby's test strategy are given. Section 4 is concerned with the alternative test strategy of Godfrey. Finally, in Section 5 some comments are given on Godfrey's test strategy. Also a third test strategy is briefly described which allows for small samples as well as for lagged dependent variables on the right hand side of the regression equation.

7.2 Thursby's test strategy

Thursby considers a regression model of the form

$$y_t = \alpha_0 + x_t'\alpha_1 + u_t, \qquad t = 1,2,\dots,n \qquad (2.1)$$

where x_t is a k×1 vector of nonstochastic regressors and u_t is an arbitrary stationary process with zero mean and variance γ_0.

Thursby's test strategy is based on a sequence of four ordered hypotheses as Godfrey (1987) remarks:

(a) explanatory variables are wrongly omitted or the functional form of the regression model is incorrect, while the disturbances follow an arbitrary but stationary ARMA process (MS);

(b) the specification of the regression model is correct and the disturbances follow an ARMA process (NAR);

(c) the specification of the regression model is correct and the disturbances follow an AR(1) process (AR(1));

(d) the specification of the regression model is correct and the disturbances are white noise (NP).

Now three consecutive tests have to be carried out. And since the hypotheses can be interpreted as increasingly restrictive assumptions the null hypotheses (b), (c) and (d) are to be tested in a sequential manner while the alternative hypotheses are (a), (b) and (c), respectively.

To test (b) against (a) Thursby proposes to use the regression specification error test (RESET). Basically the RESET test is a test of $E(u_t|x_t) = 0$. This test seems to be appropriate as Thursby (1979) has shown that the test is robust to autocorrelation. In practice the RESET test amounts to using the standard F-test whether $\delta=0$ in the regression equation

$$y_t = \alpha_0 + x_t'\alpha_1 + z_t'\delta + u_t \tag{2.2}$$

where z_t is an m-dimensional vector of test variables and δ is a vector of m unknown parameters. Various choices of variables to include in z_t have been suggested but the set found by Thursby and Schmidt (1977) to be most powerful is composed of the squares, cubes and fourth powers of the explanatory variables x_{it}. However, since there may arise a problem with the number of degrees of freedom Thursby suggests z_t should be composed of the squares and cubes or just squares of the x_{it}. An insignificant value of the F-test statistic suggests that hypothesis (b) should not be rejected.

For testing (c) against (b) Thursby follows the general approach of Sargan (1964) and Godfrey (1973). Under the null hypothesis of AR(1) disturbances $u_t=\varphi u_{t-1}+\varepsilon_t$ the well-known first order autocorrelation transformation leads to the following regression model

$$y_t = \beta_0 + x_t'\beta_1 + x_{t-1}'\beta_2 + \beta_3 y_{t-1} + \varepsilon_t \tag{2.3}$$

$$\beta_0 = \alpha_0(1-\varphi), \qquad \beta_1 = \alpha_1 \tag{2.4a}$$

$$\beta_2 = -\varphi\alpha_1, \qquad \beta_3 = \varphi \tag{2.4b}$$

Now a test of the null hypothesis of AR(1) disturbances against the alternative of NAR disturbances consists of a test of the k nonlinear restrictions implied by the AR(1) process

$$\beta_3\beta_1 + \beta_2 = 0 \tag{2.5}$$

The most obvious test of these k nonlinear restrictions is the likelihood ratio (LR) test which takes the simple form

$$LR = - n \ln(SSU/SSR) \tag{2.6}$$

LR is asymptotically $\chi^2(k)$ under the null hypothesis. SSU is the sum of squared residuals from the unrestricted estimation of (2.3) while SSR is the sum of squared residuals from the estimation of (2.3) under the restrictions (2.5). The estimation of (2.3) subject to the constraints (2.5) can readily be accomplished by using the Cochrane-Orcutt procedure or Durbin's procedure. Since there was no substantial difference in the Monte Carlo experiments Thursby proposes to use the computationally simpler procedure of Durbin which consists of the OLS estimation of

$$y_t - \tilde{\beta}_3 y_{t-1} = \beta_0 + (x_t' - \tilde{\beta}_3 x_{t-1}')\beta_1 + \varepsilon_t^* \tag{2.7}$$

where $\tilde{\beta}_3$ is the estimate of β_3 in (2.3). After estimation of (2.3) and (2.7) the statistic LR can easily be evaluated.

In his article Thursby remarks that the LR test is only valid if the disturbances of (2.3) are white noise. Hence, prior to performing the above test, a test for serial correlation among the disturbances of (2.3) must be performed. Thursby proposes to use what he calls Durbin's S-test which test takes account of the presence of y_{t-1} on the right hand side of (2.3). The test consists of the standard Student t-test for the significance of the coefficient of e_{t-1} in the regression of the OLS residuals e_t from (2.3) on the right hand side variables of (2.3) and e_{t-1}. In fact, this test is identical to the so called modified Lagrange multiplier test. Now the null hypothesis of AR(1) disturbances in (2.1) is rejected in favour of the alternative NAR if the LR test statistic is significant and/or Durbin's S-test statistic is significant. So Thursby's test of (c) against (b) is, in fact, a compound test which is referred to as the LRS procedure.

After the previous tests have not been found to be significant the well-known Durbin-Watson test statistic is used to test (d) against (c) in order to check whether the disturbances in (2.1) are white noise.

An interesting point in Thursby's work concerns the considerations on the overall significance level of his test strategy which consists of three consecutive tests, viz., RESET, LRS and DW, respectively. His Monte Carlo experiments strongly suggest that there are no dependence relationships among these tests. Hence, if α is the chosen overall significance level then the RESET, LRS and DW procedures should each be used at a significance level α^* which equals

$$\alpha^* = 1 - (1-\alpha)^{1/3} \tag{2.8}$$

Similar remarks apply to the LR test and Durbin's S-test of the compound LRS procedure.

7.3 Comments on Thursby's test strategy

7.3.1 Introduction

Monte Carlo experiments with many types of different regression models with AR(2) disturbances indicate that the test procedure proposed by Thursby can be quite powerful, particularly as sample size increases and/or as the size of the second order autocorrelation parameter increases relative to the first order parameter; see Thursby (1981). However, in spite of the suggestive Monte Carlo results Knottnerus (1985) pointed out that this conclusion does not hold for the limiting case that $\varphi_1=0$ and $\varphi_2\neq 0$, φ_1 and φ_2 being the parameters of the AR(2) process. Hence, in general, a pure seasonal autocorrelation may not be detected at all. It is also argued that Thursby's test procedure will characterize particular NAR type disturbances wrongly as AR(1) disturbances. Furthermore, it is shown that the stage in Thursby's so called LRS test where particular nonlinear restrictions on coefficients are tested is superfluous. Since these restrictions happen to be satisfied irrespective of the type of data generation process of the disturbances, the test of these restrictions is inconsistent. We shall present these criticisms now in full detail.

7.3.2 The simple AR(2) disturbances model

Consider the simple model

$$y_t = \alpha_0 + \alpha_1 x_t + u_t, \qquad t = 1,2,\dots,n \tag{3.1}$$

with true error process

$$u_t = \varphi_2 u_{t-2} + \varepsilon_t, \quad |\varphi_2| < 1 \tag{3.2}$$

where ε_t is identically and independently distributed $N(0,\sigma^2)$. Now we pay first attention to the simple Durbin-Watson test. In Thursby's test procedure the Durbin-Watson test can be performed after estimating (3.1) by ordinary least squares. Since we have $u_t=\varphi_2 u_{t-2}+\varepsilon_t$, we get $\text{plim}(\frac{1}{n}\Sigma\tilde{u}_t\tilde{u}_{t-1})=0$ for the

OLS residuals $\tilde{u}_t$. From this it follows that plim(dw)=2. So it is evident that for model (3.1) and (3.2) the value of this test statistic will generally be insignificant, if n is large enough.

In what he calls the LRS procedure, Thursby first considers the regression model

$$y_t = \beta_0 + \beta_1 x_t + \beta_2 x_{t-1} + \beta_3 y_{t-1} + w_t \tag{3.3}$$

and tests for AR(1) disturbances in this model. From (3.1) and (3.2) it can be seen that

$$\beta_0 = \alpha_0,\ \beta_1 = \alpha_1,\ \beta_2 = \beta_3 = 0 \quad \text{and} \quad w_t = u_t \tag{3.4}$$

Furthermore, it follows from (3.1), (3.2) and (3.4), that $Ey_{t-1}w_t = Eu_{t-1}u_t = 0$. Hence, the estimation of the regression model (3.3) doesn't face us with serious problems and the OLS estimators $\tilde{\beta}_i$ have

$$\text{plim}\ \tilde{\beta}_0 - \alpha_0,\ \text{plim}\ \tilde{\beta}_1 = \alpha_1,\ \text{plim}\ \tilde{\beta}_2 = \text{plim}\ \tilde{\beta}_3 = 0 \tag{3.5}$$

and by Slutzky's theorem we have $\text{plim}(\tilde{w}_t - u_t) = 0$. Because of the absence of first order autocorrelation in the true model, we may conclude that asymptotically the residuals $\tilde{w}_t$ and $\tilde{w}_{t-1}$ from (3.3) will also appear to be uncorrelated, i.e. Durbin's S-test will usually be insignificant.

The final step in the LRS test procedure is the test of the nonlinear restriction $\beta_3\beta_1 + \beta_2 = 0$. As this is true following (3.4) and as the OLS estimates are consistent in (3.3), this means that the likelihood ratio test in the test procedure of Thursby is inconsistent, i.e., the power does not tend to one when the sample size tends to infinity.

The conclusion must be that selection of the correct alternative NAR in this special case will only occur by coincidence. The test procedure of Thursby will generally select the wrong alternative of no problem with the regression, if n is large enough. Generally, any true error process with $Eu_t u_{t-1}=0$ will not be detected. This implies that Thursby's procedure is incapable of detecting pure seasonal autocorrelation, which may occur especially in monthly and quarterly data. This conclusion seems to contradict the conclusion of Thursby that the likelihood of selecting the correct alternative NAR tends to increase as the size of the second order parameter increases relative to the first order parameter.

To shed some light on these contradictory conclusions it is worthwhile to investigate the more general model of Thursby's Monte Carlo study

$$y_t = \alpha_0 + \alpha_1 x_t + u_t \tag{3.6}$$

$$u_t = \varphi_1 u_{t-1} + \varphi_2 u_{t-2} + \varepsilon_t \tag{3.7}$$

where φ_1, φ_2 and ε_t satisfy the usual conditions of an AR(2) process. To see what occurs with a test of autocorrelation in (3.3), we must examine the residuals. For simplicity we set $\alpha_0=\alpha_1=0$. Now y_t follows an AR(2) process and plim $\tilde{\beta}_3=\varphi_1/(1-\varphi_2)$, which produces residuals (asymptotically) $r_t=u_t-\gamma_1 u_{t-1}$, where $\gamma_1=\varphi_1/(1-\varphi_2)$. If we define R as the correlation coefficient between r_t and r_{t-1}, then

$$R = -\varphi_1\varphi_2/(1-\varphi_2) \tag{3.8}$$

In order to prove (3.8) define $\gamma_s=Eu_t u_{t-s}$ and set $\gamma_0=1$. It follows from the Yule-Walker equations that

$$\gamma_1 = \varphi_1/(1-\varphi_2) \tag{3.9a}$$

$$\gamma_2 = \varphi_2 + \varphi_1^2/(1-\varphi_2) \tag{3.9b}$$

Furthermore, it can easily be shown that

$$Er_t r_{t-1} = \gamma_1(\gamma_1^2 - \gamma_2) \tag{3.10a}$$

$$Er_t^2 = 1 - \gamma_1^2 \tag{3.10b}$$

and as

$$\gamma_1^2 - \gamma_2 = \gamma_1^2 - \varphi_2 - \gamma_1^2(1-\varphi_2) \tag{3.11a}$$

$$= (\gamma_1^2 - 1)\varphi_2 \tag{3.11b}$$

we have $R=-\gamma_1\varphi_2=-\varphi_1\varphi_2/(1-\varphi_2)$.

In his Monte Carlo study Thursby only considers three cases of AR(2) disturbances in model (3.6) and (3.7). The chosen values of (φ_1,φ_2) are (.8,.1), (.5,.4) and (.1,.8), respectively. The results of his Monte Carlo experiments bring Thursby to conclude that the likelihood of selecting the correct alternative NAR tends to increase as the size of the second order parameter increases relative to the first order parameter. Calculations of the absolute values of R in these three special cases give .089, .333 and .400, respectively, which are increasing indeed. However, formula (3.8) shows as well that the genuine non-monotonic character of the power function would have emerged, if Thursby had used a tighter grid, because R is a non-monotonic function of φ_2 at least if $\varphi_1=.9-\varphi_2$.

7.3.3 The general disturbances model

We now consider the general model

$$y_t = x_t'\alpha + u_t \tag{3.12}$$

where x_t and α are k×1 vectors and u_t is an arbitrary stationary process with $Eu_tu_{t-s}=\gamma_s$. For simplicity and without loss of generality we disregard the intercept term and set $\gamma_0=1$. Henceforth, the γ_s may be interpreted as correlation coefficients. In Thursby's test procedure the decisions to be taken are either NP, or AR(1) or NAR, depending on the actual values of γ_s. In performing the LRS procedure, first the model

$$y_t = x_t'\beta_1 + x_{t-1}'\beta_2 + \beta_3 y_{t-1} + w_t \tag{3.13}$$

is estimated and tested for serial correlation by Durbin's S-test. Next, the k nonlinear restrictions $\beta_3\beta_1+\beta_2=0$ are tested.

Suppose all the regressors x_{it} in (3.12) to be exogenous and write (3.13) in obvious matrix notation as

$$y = X\beta_1 + X_{-1}\beta_2 + \beta_3 y_{-1} + w \tag{3.14}$$

Next define

$$M_0 = \text{plim}\,(\tfrac{1}{n}\,X'X) \quad \text{and} \quad M_1 = \text{plim}\,(\tfrac{1}{n}\,X'X_{-1})$$

Since $y=X\alpha+u$, we obtain

$$\begin{aligned}
&\text{plim}\,(\tfrac{1}{n}\,X'y_{-1}) = M_1\alpha, && \text{plim}\,(\tfrac{1}{n}\,X_{-1}'y) = M_1'\,\alpha \\
&\text{plim}\,(\tfrac{1}{n}\,X'y) = M_0\alpha, && \text{plim}\,(\tfrac{1}{n}\,y'y) = \alpha'M_0\alpha + 1 \\
&\text{plim}\,(\tfrac{1}{n}\,y_{-1}'y) = \alpha'M_1'\alpha + \gamma_1
\end{aligned} \tag{3.15}$$

Furthermore, define the elements π_i of vector π as $\pi_i=\text{plim}(\tilde{\beta}_i)$, $\tilde{\beta}_i$ being the OLS estimator of β_i in (3.13). Then it can be seen from (3.13) and (3.15) that π_1, π_2 and π_3 have to obey the following 2k+1 asymptotic normal equations

$$\begin{bmatrix} M_0 & M_1 & M_1\alpha \\ M_1' & M_0 & M_0\alpha \\ \alpha' M_1' & \alpha' M_0 & \alpha' M_0\alpha + 1 \end{bmatrix} \begin{bmatrix} \pi_1 \\ \pi_2 \\ \pi_3 \end{bmatrix} = \begin{bmatrix} M_0\alpha \\ M_1'\alpha \\ \alpha' M_1'\alpha + \gamma_1 \end{bmatrix} \tag{3.16}$$

and it can easily be verified that

$$\pi = (\alpha', -\gamma_1\alpha', \gamma_1)' \tag{3.17}$$

Hence, we obtain $\text{plim}(\tilde{\beta}_3\tilde{\beta}_1+\tilde{\beta}_2)=0$, irrespective of the actual values γ_s, s=1,2,... . The implication is that, whatever the dependence of the disturbances, the power of the LR test will not converge to unity. An alternative and more intuitive explanation of the result (3.17) is as follows. Consider the next generalization of the well-known first order autocorrelation transformation of (3.12) where φ_1 is replaced by γ_1. This yields

$$y_t = x_t'\alpha - \gamma_1 x_{t-1}'\alpha + \gamma_1 y_{t-1} + v_t \tag{3.18}$$

$$v_t = u_t - \gamma_1 u_{t-1} \tag{3.19}$$

Then we have $Ey_{t-1}v_t=Eu_{t-1}(u_t-\gamma_1 u_{t-1})=\gamma_1-\gamma_1=0$. From this point of view OLS estimation of (3.18) or (3.13) is not a serious problem and we obtain $\text{plim}(\tilde{\beta}_3\tilde{\beta}_1+\tilde{\beta}_2)=\gamma_1\alpha-\gamma_1\alpha=0$ for any stationary process among the disturbances in (3.12).

At first glance it seems to be somewhat counterintuitive that y_{t-1} and w_t in (3.13) are independent while the disturbances w_t are serially correlated. In general, the combination of autocorrelated disturbances and the presence of y_{t-1} on the right hand side of a regression equation produces inconsistent OLS estimates because normally such a combination violates the assumption that the regressors are predetermined. However, since w_t has to be interpreted as $w_t=v_t=u_t-\gamma_1u_{t-1}$, we find $Ey_{t-1}w_t=0$.

It emerges that the test of the nonlinear restrictions is superfluous. The test doesn't provide any information whatsoever on the type of disturbances in (3.12). It is only Durbin's S-test in (3.13), in combination with the result of the Durbin and Watson statistic in (3.12), which are used by Thursby's test procedure to discriminate between NAR, AR(1) and NP in (3.12).

Because of (3.17) the residuals $\tilde{w}_t$ from (3.13) obey $\text{plim}(\tilde{w}_t)=u_t-\gamma_1u_{t-1}$. Hence,

$$\text{plim}\left(\frac{1}{n}\Sigma\,\tilde{w}_t\tilde{w}_{t-1}\right) = \gamma_1(\gamma_1^2 - \gamma_2) \qquad (3.20)$$

If $\gamma_2=\gamma_1^2$ or $\gamma_1=0$ this expression equals zero, and then Durbin's S-test for autocorrelation will usually be insignificant. Hence, on the one hand Thursby's LRS procedure will wrongly characterize NAR type disturbances as AR(1) disturbances if $\gamma_2=\gamma_1^2\neq0$. On the other hand NAR type disturbances with $\gamma_1=0$ may, in general, not be detected at all.

7.4 Godfrey's test strategy

Referring to Porter and Kashyap (1984) Godfrey (1987) remarks that the RESET test is sensitive to a lack of independence of the disturbances u_t, particularly if the regressors themselves are serially correlated. From the Monte Carlo experiments of Porter and Kashyap it emerges that autocorrelation among the disturbances can lead to a high number of false detections of misspecification (MS) by the RESET test. This is in line with the finding of Palm and Sneek (1984) that the F-test statistic can substantially be affected by autocorrelation among the disturbances. The implication is that the RESET test is not an appropriate test to test hypothesis (b) against (a).

Thursby's assumption that the RESET test is robust to autocorrelation of the disturbances is based upon some Monte Carlo experiments in which the regressors were serially uncorrelated; see Thursby (1979) for further details. However, this is rarely the case in the context of economic time series.

Godfrey proposes to use a test based on the work of White and Domowitz (1984) which is capable of detecting misspecification while it is robust to autocorrelation among the disturbances. To describe the test it is convenient to write (2.2) in matrix notation

$$y = W\Theta + u \tag{4.1}$$

where $W=(X,Z)$, $\Theta'=(\alpha',\delta')$ and Z is the $n\times m$ matrix of test variables. Now a robust test of $\delta=0$ is as follows. The restrictions $\delta=0$ can be written in the form $R\Theta=0$, R being equal to $R=(0,I_m)$. If V denotes the covariance matrix of the OLS estimator $\tilde{\Theta}$ then it can easily be shown that V is equal to

$$V = (W'W)^{-1}\ W'\Gamma W\ (W'W)^{-1} \tag{4.2}$$

where Γ denotes the covariance matrix of the disturbances u_t. Hence, an asymptotically valid test statistic is

$$g = (R\tilde{\Theta})'(R\tilde{V}R')^{-1}(R\tilde{\Theta}) \tag{4.3}$$

which is asymptotically $\chi^2(m)$ under the null hypothesis $\delta=0$. Here $\tilde{V}$ equals V except that Γ in (4.2) is replaced by its estimate $\tilde{\Gamma}$.

There are several methods to estimate Γ. White and Domowitz suggest setting

$$\tilde{\Gamma}_{ts} = \tilde{u}_t\tilde{u}_s \quad \text{if } |t-s| < j \tag{4.4a}$$

$$\tilde{\Gamma}_{ts} = 0 \quad \text{if } |t-s| \geqq j \tag{4.4b}$$

where $\tilde{u}_t$ is the OLS residual from (2.1) and j is a truncation parameter to be chosen more or less arbitrarily. The choice of j should at least take account of the seasonal pattern of the data, i.e., j>4 in the case of quarterly data, etc. White and Domowitz propose to base the choice of j on regressions of $\tilde{u}_t\tilde{u}_{t-h}$ on terms of $x_t x'_{t-h}$, h=1,2,... . It should be noted that the matrix estimator (4.4) is also heteroskedasticity consistent; see Godfrey (1987, p. 131).

An alternative procedure for estimating Γ was proposed by a referee of Godfrey's article. He suggests to fit an appropriate ARMA model to the residuals $\tilde{u}_t$. Next the autocovariances can be calculated; see Chapter 2 or see Box and Jenkins (1976, Chapter 3). This estimator of Γ might be superior to the estimator proposed by White and Domowitz.

The second step in Godfrey's test procedure consists of a test of AR(1) disturbances against the alternative hypothesis of an arbitrary stationary process among the disturbances. If the alternative hypothesis is replaced by the more specific alternative of a particular AR(1+q) process, then the classical and asymptotically equivalent tests, viz., the Lagrange multiplier (LM) test, the Wald (W) test and the likelihood ratio (LR) test are available. In the models considered by Thursby even a very simple test is available. Since by assumption all explanatory variables are exogenous it can be shown that the classical tests are asymptotically equivalent to the standard F-test of $\varphi_2=\varphi_3=\ldots=\varphi_{1+q}=0$ in the regression

$$\tilde{u}_t = \varphi_1\tilde{u}_{t-1} + \ldots + \varphi_{1+q}\tilde{u}_{t-q-1} + a_t \tag{4.5}$$

a_t being an error term. For a proof see Godfrey (1987). Furthermore, Godfrey (1981) pointed out that the same test can be used as a test of an AR(1) process against the alternative of an ARMA(1,q) process. To detect pure seasonal autocorrelation it is advised to take q=3 or 4 in the case of quarterly data and q=11 or 12 in the case of monthly data.

If the test statistic from (4.5) is insignificant then in the third step of his test procedure Godfrey proposes like Thursby in the final stage of his test sequence, to use the well-known Durbin-Watson test in order to discriminate between the alternatives (c) and (d). An asymptotically equivalent test is the test of $\varphi_1=0$ in

$$\tilde{u}_t = \varphi_1 \tilde{u}_{t-1} + a_t \tag{4.6}$$

Finally, Godfrey pays some attention to cases where one of the explanatory variables is a lagged dependent variable. In that case all the tests of his test procedure are invalid. Given an AR(1+q) or an ARMA(1,q) process among the disturbances Godfrey suggests estimating (2.1) and (2.2) by a suitable maximum likelihood procedure. Then the restrictions $\delta=0$ in (2.2) can be tested by means of the LR test. If this LR test statistic is insignificant then (2.1) should also be estimated under the assumption of AR(1) disturbances in order to apply once more the LR test but now for discriminating between the hypotheses (b) and (c). Finally, if (c) is not rejected, Durbin's h-test can be used for testing (d), the hypothesis that there is no problem with model (2.1).

7.5 Comments on Godfrey's test strategy

A quite general disadvantage of Godfrey's test procedure is that apart from the Durbin-Watson test, all tests in Godfrey's strategy are asymptotically valid only. Besides Durbin's h-test statistic is of little help in the case that it takes an imaginary value; note that the latter problem can easily be overcome by using Durbin's S-test. In the sequel of this section we propose some minor changes in Godfrey's test procedure. These allow for the possible presence of a lagged dependent variable and entail modifications which hopefully improve the performance of the strategy in small samples.

It is recalled that Godfrey's test statistic g from (4.3) is robust to serially correlated disturbances and that it is asymptotically valid, whereas the RESET test in Thursby's strategy is an exact test. Unfortunately, the problem of the RESET test is that it suffers from a lack of robustness in case of serially correlated disturbances as Porter and Kashyap pointed out.

Here we suggest an alternative test. It is based on both the RESET test and Godfrey's test; it discriminates between (a) and (b) and is robust to autocorrelation. At the same time it takes into account the small sample character of most time series. The procedure is as follows. Calculate the OLS residuals from (2.1) and fit an appropriate ARMA model to these residuals. Then consider the GLS transformation of (2.1)

$$Py = PX\alpha + Pu \tag{5.1}$$

The transformed disturbances of this equation obey the classical assumptions and the RESET test can be applied. In the case of presence of lagged dependent variables on the right hand side of (2.1) this approach can be followed as well. In a first step the parameters of (2.1) should be estimated in a consistent manner by means of one of the estimation methods described in the previous chapter. Then calculate in a second step the corresponding residuals, fit an ARMA model to these residuals and apply the RESET test to (5.1), i.e., the GLS transformation of (2.1); in the Chapters 2 and 3 it is explained how such a transformation can be carried out.

In order to discriminate between the hypotheses (b), (c) and (d) it seems to be worthwhile to employ the modified Lagrange multiplier (MLM) test procedure. The MLM test for discriminating between (b) and (c) consists of a standard F-test of $\varphi_2=\ldots=\varphi_{1+q}=0$ in the regression of the residuals $\tilde{u}_t$ from (2.1) on the right-hand side variables of (2.1) and $\tilde{u}_{t-1},\ldots,\tilde{u}_{t-q-1}$. Note that Godfrey uses (4.5) wherein the right hand side variables from (2.1) are omitted. Kiviet (1986) pointed out that an advantage of the MLM test is that its actual significance level closely corresponds to the nominal significance level in small samples. Moreover, the MLM test is also valid in the case that one of the explanatory variables of (2.1) is a lagged dependent variable, provided that (2.1) is estimated under the null in a consistent manner. Furthermore, the MLM test seems to be computationally more attractive than the LR test suggested by Godfrey at least in the case of a lagged dependent variable on the right hand side of (2.1); note that if the LR test is applied (2.1) has to be estimated under the alternative hypothesis too. Finally note that in the case of AR(1) disturbances the MLM test is identical with Durbin's S-test.

REFERENCES

ABRAHAM, B. and J. LEDOLTER (1983), Statistical Methods for Forecasting, John Wiley & Sons, New York.

AMEMIYA, T. (1973), "Generalized Least Squares with an Estimated Autocovariance Matrix," Econometrica, 41, 723-732.

AMEMIYA, T. and W.A. FULLER (1967), "A Comparative Study of Alternative Estimators in a Distributed Lag Model," Econometrica, 35, 509-529.

ANDERSON, B.D.O. and J.B. MOORE (1979), Optimal Filtering, Prentice-Hall, Englewood Cliffs, N.J.

ANSLEY, C.F. (1979), "An Algorithm for the Exact Likelihood of a Mixed Autoregressive-Moving Average Process," Biometrika, 66, 59-65.

ANSLEY, C.F. (1980), "Computation of the Theoretical Autocovariance Function for a Vector ARMA Process," Journal of Statistical Computation and Simulation, 12, 15-24.

ARATO, M. (1961), "On the Sufficient Statistics for Stationary Gaussian Random Processes," Theoretical Probability and its Applications, 6, 199-201.

ÅSTRÖM, K.J. (1970), Introduction to Stochastic Control Theory, Academic Press, New York.

BALESTRA, P. (1980), "A Note on the Exact Transformation Associated with the First Order Moving Average Process," Journal of Econometrics, 14, 381-394.

BEACH, C.M. and J.G. MACKINNON (1978), "A Maximum Likelihood Procedure for Regression with Autocorrelated Errors," Econometrica, 46, 51-58.

BERTSEKAS, D.P. (1976), Dynamic Programming and Control, Academic Press, New York.

BOX, G.E.P. and G.M. JENKINS (1976), Time Series Analysis: Forecasting and Control, Holden Day, San Francisco.

CHOW, G.C. (1983), Econometrics, McGraw-Hill, Tokyo.

CRAMER, J.S. (1986), Econometric Applications of Maximum Likelihood Methods, Cambridge University Press, Cambridge.

DHRYMES, P.J. (1969), "Efficient Estimation of Distributed Lags with Autocorrelated Errors," International Economic Review, 10, 47-67.

DUNCAN, D.B. and S.D. HORN (1972), "Linear Dynamic Recursive Estimation from the Viewpoint of Regression Analysis," Journal of the American Statistical Association, 67, 815-821.

DURBIN, J. (1960), "The Fitting of Time Series Models," Review of the International Statistical Institute, 28, 233-244.

FRIEDMAN, A. (1982), Foundations of Modern Analysis, Dover, New York.

GODFREY, L.G. (1973), "A Note on the Treatment of Serial Correlation," Canadian Journal of Economics, 6, 567-573.

GODFREY, L.G. (1981), "On the Invariance of the Lagrange Multiplier Test with Respect to Certain Changes in the Alternative Hypothesis," Econometrica, 49, 1303-1310.

GODFREY, L.G. (1987), "Discriminating between Autocorrelation and Misspecification in Regression Analysis: An Alternative Test Strategy," The Review of Economics and Statistics, 69, 128-134.

GOLDBERG, S. (1958), Difference Equations, John Wiley, New York.

GOODWIN, G.C. and K.S. SIN (1984), Adaptive Filtering Prediction and Control, Prentice-Hall, Englewood Cliffs, N.J.

GOOIJER, J.G. DE (1978), "On the Inverse of the Autocovariance Matrix for a General Mixed Autoregressive Moving Average Process," Statistische Hefte, 19, 114-123.

GOOIJER, J.G. DE (1984), Contributions to Univariate Time Series Analysis with an Application to Dutch Stock Market Prices, Pasmans, The Hague.

GUPTA, Y.P. (1971), Statistical Estimation of Linear Economic Relationships, Rotterdam University Press, Rotterdam.

HAMILTON, J.D. (1986), "A Standard Error for the Estimated State Vector of a State-Space Model," Journal of Econometrics, 33, 387-397.

HANNAN, E.J. and M. DEISTLER (1988), The Statistical Theory of Linear Systems, Wiley, New York.

HARVEY, A.C. (1981), Time Series Models, Phillip Allan, Deddington.

HARVEY, A.C. (1986), The Econometric Analysis of Time Series, Phillip Allan, Deddington.

HARVEY, A.C. and I.D. MCAVINCHEY (1978), The Small Sample Efficiency of Two-step Estimators in Regression Models with Autoregressive Disturbances, Discussion Paper No. 78-10, University of British Columbia.

HARVEY, A.C. and G.D.A. PHILLIPS (1979), "Maximum Likelihood Estimation of Regression Models with Autoregressive-Moving Average Disturbances," Biometrika, 66, 49-58.

HARVEY, A.C. and R.G. PIERSE (1984), "Estimating Missing Observations in Economic Time Series," Journal of the American Statistical Association, 79, 125-131.

HATANAKA, M. (1974), "An Efficient Two Step Estimator for the Dynamic Adjustment Model with Autoregressive Errors," Journal of Econometrics, 2, 199-200.

JONES, R.H. (1980), "Maximum Likelihood Fitting of ARMA Models to Time Series with Missing Observations," Technometrics, 22, 389-395.

JUDGE, G.G., W.E. GRIFFITHS, R.C. HILL and T.C. LEE (1980), The Theory and Practice of Econometrics, John Wiley & Sons, New York.

KAILATH, T. (1981), Lectures on Wiener and Kalman Filtering, Springer Verlag, Wien.

KALMAN, R.E. (1960), "A New Approach to Linear Filtering and Prediction Problems," Transactions ASME, Journal of Basic Engineering, 82, 35-45.

KIVIET, J.F. (1986), "On the Rigour of Some Misspecification Tests for Modelling Dynamic Relationships," Review of Economic Studies, 53, 241-261.

KLEIN, L.R. (1958), "The Estimation of Distributed Lags," Econometrica, 26, 553-565.

KNOTTNERUS, P. (1985), "A Test Strategy for Discriminating between Autocorrelation and Misspecification in Regression Analysis: A Critical Note," The Review of Economics and Statistics, 67, 175-177.

KOHN, R. and C.F. ANSLEY (1982), "A Note on Obtaining the Theoretical Autocovariances of an ARMA Process," Journal of Statistical Computation and Simulation, 15, 273-283.

KOYCK, L.M. (1954), Distributed Lags and Investment Analysis, North-Holland, Amsterdam.

LEVINSON, N. (1946), "The Wiener RMS (root mean square) Error Criterion in Filter Design and Prediction," Journal of Mathematical Physics, 25, 81-94.

LIVIATAN, N. (1963), "Consistent Estimation of Distributed Lags," International Economic Review, 4, 44-52.

LJUNG, G.M. and G.E.P. BOX (1979), "The Likelihood Function of Stationary Autoregressive-Moving Average Models," Biometrika, 66, 265-270.

MAGNUS, J.R. (1978), "Maximum Likelihood Estimation of the GLS Model with Unknown Parameters in the Disturbance Covariance Matrix," Journal of Econometrics, 7, 281-312.

MAGNUS, J.R. and H. NEUDECKER (1988), Matrix Differential Calculus with Applications in Statistics and Econometrics, John Wiley & Sons, Chichester.

MCLEOD, A.I. (1975, 1977), "Derivation of the Theoretical Autocovariance Function of Autoregressive-Moving Average Time Series," Applied Statistics, 24, 255-256. Correction, 26, 194.

MÉLARD, G. (1983), "Algorithm AS 197: A Fast Algorithm for the Exact Likelihood of Autoregressive-Moving Average Models," Applied Statistics, 32, 104-114.

MENTZ, R.P. (1976), "On the Inverse of Some Covariance Matrices of Toeplitz Type," Journal of Applied Mathematics, 31, 426-437.

NERLOVE, M. (1958), _The Dynamics of Supply: Estimation of Farmers' Response to Price_, John Hopkins Press, Baltimore.

NERLOVE, M., D.M. GRETHER and J.L. CARVALHO (1979), _Analysis of Economic Time Series_, Academic Press, New York.

OBERHOFER, W. and J. KMENTA (1974), "A General Procedure for Obtaining Maximum Likelihood Estimates in Generalized Regression Models," _Econometrica_, 42, 579-590.

PALM, F.C. and J.M. SNEEK (1984), "Significance Tests and Spurious Correlation in Regression Models with Autocorrelated Errors," _Statistische Hefte_, 25, 87-105.

PESARAN, M.H. (1973), "Exact Maximum Likelihood Estimation of a Regression Equation with a First Order Moving Average Error," _Review of Economic Studies_, 40, 529-536.

PHADKE, M.S. and G. KEDEM (1978), "Computation of the Exact Likelihood Function of Multivariate Moving Average Models," _Biometrika_, 65, 511-519.

PHILLIPS, G.D.A. and A.C. HARVEY (1974), "A Simple Test for Serial Correlation in Regression Analysis," _Journal of the American Statistical Association_, 69, 935-939.

PORTER, R.D. and A.K. KASHYAP (1984), "Autocorrelation and the Sensitivity of RESET," _Economics Letters_, 14, 229-233.

RICHARDSON, S.M. and K.J. WHITE (1979), "The Power of Tests for Autocorrelation with Missing Observations," _Econometrica_, 47, 785-788.

SARGAN, J.D. (1964), "Wages and Prices in the United Kingdom: A Study in Econometric Methodology," in P.E. Hart et al. (eds.), _Econometric Analysis for National Economic Planning_ (The Colston Research Society), 25-55.

SAVIN, N.E. and K.J. WHITE (1978a), "Testing for Autocorrelation with Missing Observations," _Econometrica_, 46, 59-67.

SAVIN, N.E. and K.J. WHITE (1978b), "Estimation and Testing for Functional Form and Autocorrelation: A Simultaneous Approach," Journal of Econometrics, 8, 1-12.

SCHMIDT, P. (1975), "The Small Sample Effects of Various Treatments of Truncation Remainders in the Estimation of Distributed Lag Models," Review of Economics and Statistics," 57, 387-389.

SHAMAN, P. (1975), "An Approximate Inverse for the Covariance Matrix of Moving Average and Autoregressive Processes," The Annals of Statistics, 3, 532-538.

SHILOV, G.E. (1977), Linear Algebra, Dover, New York.

SPANOS, A. (1986), Statistical Foundations of Econometric Modelling, Cambridge University Press, Cambridge.

STEYN, I.J. (1988), "Starting up the Kalman Filter: Options and Consequences," Paper presented at Ecozoek Dag 1988, Maastricht.

STOER, J. (1976), Einführung in die Numerische Mathematik I, Springer Verlag, Heidelberg.

THEIL, H. (1971), Principles of Econometrics, John Wiley & Sons, New York.

THURSBY, J.G. (1979), "Alternative Specification Error Tests: A Comparative Study," Journal of the American Statistical Association, 74, 222-225.

THURSBY, J.G. (1981), "A Test Strategy for Discriminating between Autocorrelation and Misspecification in Regression Analysis," The Review of Economics and Statistics, 63, 117-123.

THURSBY, J.G. (1985), "A Test Strategy for Discriminating between Autocorrelation and Misspecification in Regression Analysis: A Reply," The Review of Economics and Statistics, 67, 177-178.

THURSBY, J.G. and P. SCHMIDT (1977), "Some Properties of Tests for Specification Error in a Linear Regression Model," Journal of the American Statistical Association, 72, 635-641.

TIAO, G.C. and M.M. ALI (1971), "Analysis of Correlated Random Effects: Linear Model with Two Random Components," Biometrika, 58, 37–51.

WANSBEEK, T.J. and A. KAPTEYN (1985), "Estimation in a Linear Model with Serially Correlated Errors when Observations are Missing," International Economic Review, 26, 469–490.

WHITE, H. and I. DOMOWITZ (1984), "Nonlinear Regression with Dependent Observations," Econometrica, 52, 143-161.

ZINDE-WALSH, V. (1988), "Some Exact Formulae for Autoregressive Moving Average Processes," Econometric Theory, 4, 384–402.

AUTHOR INDEX

SUBJECT INDEX

Vol. 261: Th.R. Gulledge, Jr., N.K. Womer, The Economics of Made-to-Order Production. VI, 134 pages. 1986.

Vol. 262: H.U. Buhl, A Neo-Classical Theory of Distribution and Wealth. V, 146 pages. 1986.

Vol. 263: M. Schäfer, Resource Extraction and Market Structure. XI, 154 pages. 1986.

Vol. 264: Models of Economic Dynamics. Proceedings, 1983. Edited by H.F. Sonnenschein. VII, 212 pages. 1986.

Vol. 265: Dynamic Games and Applications in Economics. Edited by T. Başar. IX, 288 pages. 1986.

Vol. 266: Multi-Stage Production Planning and Inventory Control. Edited by S. Axsäter, Ch. Schneeweiss and E. Silver. V, 264 pages. 1986.

Vol. 267: R. Bemelmans, The Capacity Aspect of Inventories. IX, 165 pages. 1986.

Vol. 268: V. Firchau, Information Evaluation in Capital Markets. VII, 103 pages. 1986.

Vol. 269: A. Borglin, H. Keiding, Optimality in Infinite Horizon Economies. VI, 180 pages. 1986.

Vol. 270: Technological Change, Employment and Spatial Dynamics. Proceedings 1985. Edited by P. Nijkamp. VII, 466 pages. 1986.

Vol. 271: C. Hildreth, The Cowles Commission in Chicago, 1939–1955. V, 176 pages. 1986.

Vol. 272: G. Clemenz, Credit Markets with Asymmetric Information. VIII, 212 pages. 1986.

Vol. 273: Large-Scale Modelling and Interactive Decision Analysis. Proceedings, 1985. Edited by G. Fandel, M. Grauer, A. Kurzhanski and A.P. Wierzbicki. VII, 363 pages. 1986.

Vol. 274: W.K. Klein Haneveld, Duality in Stochastic Linear and Dynamic Programming. VII, 295 pages. 1986.

Vol. 275: Competition, Instability, and Nonlinear Cycles. Proceedings, 1985. Edited by W. Semmler. XII, 340 pages. 1986.

Vol. 276: M.R. Baye, D.A. Black, Consumer Behavior, Cost of Living Measures, and the Income Tax. VII, 119 pages. 1986.

Vol. 277: Studies in Austrian Capital Theory, Investment and Time. Edited by M. Faber. VI, 317 pages. 1986.

Vol. 278: W.E. Diewert, The Measurement of the Economic Benefits of Infrastructure Services. V, 202 pages. 1986.

Vol. 279: H.-J. Büttler, G. Frei and B. Schips, Estimation of Disequilibrium Models. VI, 114 pages. 1986.

Vol. 280: H.T. Lau, Combinatorial Heuristic Algorithms with FORTRAN. VII, 126 pages. 1986.

Vol. 281: Ch.-L. Hwang, M.-J. Lin, Group Decision Making under Multiple Criteria. XI, 400 pages. 1987.

Vol. 282: K. Schittkowski, More Test Examples for Nonlinear Programming Codes. V, 261 pages. 1987.

Vol. 283: G. Gabisch, H.-W. Lorenz, Business Cycle Theory. VII, 229 pages. 1987.

Vol. 284: H. Lütkepohl, Forecasting Aggregated Vector ARMA Processes. X, 323 pages. 1987.

Vol. 285: Toward Interactive and Intelligent Decision Support Systems. Volume 1. Proceedings, 1986. Edited by Y. Sawaragi, K. Inoue and H. Nakayama. XII, 445 pages. 1987.

Vol. 286: Toward Interactive and Intelligent Decision Support Systems. Volume 2. Proceedings, 1986. Edited by Y. Sawaragi, K. Inoue and H. Nakayama. XII, 450 pages. 1987.

Vol. 287: Dynamical Systems. Proceedings, 1985. Edited by A.B. Kurzhanski and K. Sigmund. VI, 215 pages. 1987.

Vol. 288: G.D. Rudebusch, The Estimation of Macroeconomic Disequilibrium Models with Regime Classification Information. VII, 128 pages. 1987.

Vol. 289: B.R. Meijboom, Planning in Decentralized Firms. X, 168 pages. 1987.

Vol. 290: D.A. Carlson, A. Haurie, Infinite Horizon Optimal Control. XI, 254 pages. 1987.

Vol. 291: N. Takahashi, Design of Adaptive Organizations. VI, 140 pages. 1987.

Vol. 292: I. Tchijov, L. Tomaszewicz (Eds.), Input-Output Modeling. Proceedings, 1985. VI, 195 pages. 1987.

Vol. 293: D. Batten, J. Casti, B. Johansson (Eds.), Economic Evolution and Structural Adjustment. Proceedings, 1985. VI, 382 pages. 1987.

Vol. 294: J. Jahn, W. Krabs (Eds.), Recent Advances and Historical Development of Vector Optimization. VII, 405 pages. 1987.

Vol. 295: H. Meister, The Purification Problem for Constrained Games with Incomplete Information. X, 127 pages. 1987.

Vol. 296: A. Börsch-Supan, Econometric Analysis of Discrete Choice. VIII, 211 pages. 1987.

Vol. 297: V. Fedorov, H. Läuter (Eds.), Model-Oriented Data Analysis. Proceedings, 1987. VI, 239 pages. 1988.

Vol. 298: S.H. Chew, Q. Zheng, Integral Global Optimization. VII, 179 pages. 1988.

Vol. 299: K. Marti, Descent Directions and Efficient Solutions in Discretely Distributed Stochastic Programs. XIV, 178 pages. 1988.

Vol. 300: U. Derigs, Programming in Networks and Graphs. XI, 315 pages. 1988.

Vol. 301: J. Kacprzyk, M. Roubens (Eds.), Non-Conventional Preference Relations in Decision Making. VII, 155 pages. 1988.

Vol. 302: H.A. Eiselt, G. Pederzoli (Eds.), Advances in Optimization and Control. Proceedings, 1986. VIII, 372 pages. 1988.

Vol. 303: F.X. Diebold, Empirical Modeling of Exchange Rate Dynamics. VII, 143 pages. 1988.

Vol. 304: A. Kurzhanski, K. Neumann, D. Pallaschke (Eds.), Optimization, Parallel Processing and Applications. Proceedings, 1987. VI, 292 pages. 1988.

Vol. 305: G.-J.C.Th. van Schijndel, Dynamic Firm and Investor Behaviour under Progressive Personal Taxation. X, 215 pages. 1988.

Vol. 306: Ch. Klein, A Static Microeconomic Model of Pure Competition. VIII, 139 pages. 1988.

Vol. 307: T.K. Dijkstra (Ed.), On Model Uncertainty and its Statistical Implications. VII, 138 pages. 1988.

Vol. 308: J.R. Daduna, A. Wren (Eds.), Computer-Aided Transit Scheduling. VIII, 330 pages. 1988.

Vol. 309: G. Ricci, K. Velupillai (Eds.), Growth Cycles and Multisectoral Economics: the Goodwin Tradition. III, 126 pages. 1988.

Vol. 310: J. Kacprzyk, M. Fedrizzi (Eds.), Combining Fuzzy Imprecision with Probabilistic Uncertainty in Decision Making. IX, 399 pages. 1988.

Vol. 311: R. Färe, Fundamentals of Production Theory. IX, 163 pages. 1988.

Vol. 312: J. Krishnakumar, Estimation of Simultaneous Equation Models with Error Components Structure. X, 357 pages. 1988.

Vol. 313: W. Jammernegg, Sequential Binary Investment Decisions. VI, 156 pages. 1988.

Vol. 314: R. Tietz, W. Albers, R. Selten (Eds.), Bounded Rational Behavior in Experimental Games and Markets. VI, 308 pages. 1988.

Vol. 315: I. Orishimo, G.J.D. Hewings, P. Nijkamp (Eds.), Information Technology: Social and Spatial Perspectives. Proceedings, 1986. VI, 268 pages. 1988.

Vol. 316: R.L. Basmann, D.J. Slottje, K. Hayes, J.D. Johnson, D.J. Molina, The Generalized Fechner-Thurstone Direct Utility Function and Some of its Uses. VIII, 159 pages. 1988.

Vol. 317: L. Bianco, A. La Bella (Eds.), Freight Transport Planning and Logistics. Proceedings, 1987. X, 568 pages. 1988.

Vol. 318: T. Doup, Simplicial Algorithms on the Simplotope. VIII, 262 pages. 1988.

Vol. 319: D.T. Luc, Theory of Vector Optimization. VIII, 173 pages. 1989.

Vol. 320: D. van der Wijst, Financial Structure in Small Business. VII, 181 pages. 1989.

Vol. 321: M. Di Matteo, R.M. Goodwin, A. Vercelli (Eds.), Technological and Social Factors in Long Term Fluctuations. Proceedings. IX, 442 pages. 1989.

Vol. 322: T. Kollintzas (Ed.), The Rational Expectations Equilibrium Inventory Model. XI, 269 pages. 1989.

Vol. 323: M.B.M. de Koster, Capacity Oriented Analysis and Design of Production Systems. XII, 245 pages. 1989.

Vol. 324: I.M. Bomze, B.M. Pötscher, Game Theoretical Foundations of Evolutionary Stability. VI, 145 pages. 1989.

Vol. 325: P. Ferri, E. Greenberg, The Labor Market and Business Cycle Theories. X, 183 pages. 1989.

Vol. 326: Ch. Sauer, Alternative Theories of Output, Unemployment, and Inflation in Germany: 1960–1985. XIII, 206 pages. 1989.

Vol. 327: M. Tawada, Production Structure and International Trade. V, 132 pages. 1989.

Vol. 328: W. Güth, B. Kalkofen, Unique Solutions for Strategic Games. VII, 200 pages. 1989.

Vol. 329: G. Tillmann, Equity, Incentives, and Taxation. VI, 132 pages. 1989.

Vol. 330: P.M. Kort, Optimal Dynamic Investment Policies of a Value Maximizing Firm. VII, 185 pages. 1989.

Vol. 331: A. Lewandowski, A.P. Wierzbicki (Eds.), Aspiration Based Decision Support Systems. X, 400 pages. 1989.

Vol. 332: T.R. Gulledge, Jr., L.A. Litteral (Eds.), Cost Analysis Applications of Economics and Operations Research. Proceedings. VII, 422 pages. 1989.

Vol. 333: N. Dellaert, Production to Order. VII, 158 pages. 1989.

Vol. 334: H.-W. Lorenz, Nonlinear Dynamical Economics and Chaotic Motion. XI, 248 pages. 1989.

Vol. 335: A. G. Lockett, G. Islei (Eds.), Improving Decision Making in Organisations. Proceedings. IX, 606 pages. 1989.

Vol. 336: T. Puu, Nonlinear Economic Dynamics. VII, 119 pages. 1989.

Vol. 337: A. Lewandowski, I. Stanchev (Eds.), Methodology and Software for Interactive Decision Support. VIII, 309 pages. 1989.

Vol. 338: J.K. Ho, R.P. Sundarraj, DECOMP: an Implementation of Dantzig-Wolfe Decomposition for Linear Programming. VI, 206 pages. 1989.

Vol. 339: J. Terceiro Lomba, Estimation of Dynamic Econometric Models with Errors in Variables. VIII, 116 pages. 1990.

Vol. 340: T. Vasko, R. Ayres, L. Fontvieille (Eds.), Life Cycles and Long Waves. XIV, 293 pages. 1990.

Vol. 341: G.R. Uhlich, Descriptive Theories of Bargaining. IX, 165 pages. 1990.

Vol. 342: K. Okuguchi, F. Szidarovszky, The Theory of Oligopoly with Multi-Product Firms. V, 167 pages. 1990.

Vol. 343: C. Chiarella, The Elements of a Nonlinear Theory of Economic Dynamics. IX, 149 pages. 1990.

Vol. 344: K. Neumann, Stochastic Project Networks. XI, 237 pages. 1990.

Vol. 345: A. Cambini, E. Castagnoli, L. Martein, P. Mazzoleni, S. Schaible (Eds.), Generalized Convexity and Fractional Programming with Economic Applications. Proceedings, 1988. VII, 361 pages. 1990.

Vol. 346: R. von Randow (Ed.), Integer Programming and Related Areas. A Classified Bibliography 1984–1987. XIII, 514 pages. 1990.

Vol. 347: D. Ríos Insua, Sensitivity Analysis in Multi-objective Decision Making. XI, 193 pages. 1990.

Vol. 348: H. Störmer, Binary Functions and their Applications. VIII, 151 pages. 1990.

Vol. 349: G. A. Pfann, Dynamic Modelling of Stochastic Demand for Manufacturing Employment. VI, 158 pages. 1990.

Vol. 350: W.-B. Zhang, Economic Dynamics. X, 232 pages. 1990.

Vol. 351: A. Lewandowski, V. Volkovich (Eds.), Multiobjective Problems of Mathematical Programming. Proceedings, 1988. VII, 315 pages. 1991.

Vol. 352: O. van Hilten, Optimal Firm Behaviour in the Context of Technological Progress and a Business Cycle. XII, 229 pages. 1991.

Vol. 353: G. Ricci (Ed.), Decision Processes in Economics. Proceedings, 1989. III, 209 pages. 1991.

Vol. 354: M. Ivaldi, A Structural Analysis of Expectation Formation. XII, 230 pages. 1991.

Vol. 355: M. Salomon, Deterministic Lotsizing Models for Production Planning. VII, 158 pages. 1991.

Vol. 356: P. Korhonen, A. Lewandowski, J. Wallenius (Eds.), Multiple Criteria Decision Support. Proceedings, 1989. XII, 393 pages. 1991.

Vol. 358: P. Knottnerus, Linear Models with Correlated Disturbances. VIII, 196 pages. 1991.